CONTENTS

Title Page
Copyright
Dedication
Introduction: Stardust and Breath
A Prelude to Wonder

Chapter 1: The Quantum Bridge of Wonder	1
Chapter 2: The Fabric of Reality	8
Chapter 3: The Quantum Bridge	15
Chapter 4: The Divine Algorithm	36
Chapter 5: Entanglement	52
Chapter 6: The Nature of Time and Timelessness	63
Chapter 7: The Imprint of the Divine	75
Chapter 8: Humanity's Place in the Web of Life	87
Chapter 9: The Quantum Mind of God	104
A Letter to the Future of Humanity:	121

THE QUANTUM MIND OF GOD

The Resonance of Creation Unfolding

Ben Maestas

Copyright © 2024 Ben Maestas

All rights reserved

Printed in the United States of America

*Thank you to Aimeé, my beloved, I could not have grown to this point, let alone been able to devote so much time to this undertaking without her support, above and beyond all measure. I am also very grateful to the collective of minds that went into the writing of this book. From the thinkers to the believers and even the observers, all of you played a part in the success of this endeavor.
A special shout-out to Aaron Illimité for the countless hours of being my sounding board, biggest critic, editor, and friend! Thank you all for being you. This book is a testament to what unity can create.*

INTRODUCTION: STARDUST AND BREATH

Somewhere in the vast reaches of space billions of years ago, a star was dying. Its final act was one of both destruction and creation, a cataclysmic explosion that scattered its very essence across the universe. This was the beginning of everything we know, a moment when the cosmos became the forge for the elements that would one day shape us. Among the remnants of that stellar death was iron, flung across the void, waiting for its time to shine again.

The universe moved with purpose. Gravity gathered these scattered pieces into stars, galaxies, and eventually the Milky Way, where another star was born, our Sun. Around it swirled a disk of gas and dust, coalescing into planets, including one that would be unique: Earth. A collision with a Mars-sized object gave Earth its Moon, stabilizing its tilt and setting the stage for seasons, tides, and long-term climate stability.

As Earth cooled water began to pool, delivered by asteroids and comets during the Late Heavy Bombardment. Within these oceans, a miracle occurred: molecules arranged themselves into self-replicating systems. Life began, humbly but relentlessly, transforming the world. Cyanobacteria's gift of photosynthesis filled the skies with oxygen, and complex cells emerged, carrying tiny symbiotic partners we now call mitochondria. These cells, the foundation of all complex life, were the first step toward the incredible diversity we see today.

Millions of years later, a new kind of creature would emerge. Walking upright, creating tools, mastering fire, and eventually speaking, humanity stood apart. We were not just observers of the world, we were participants in its transformation, bearers of a unique consciousness capable of reflection and co-creation.

Every atom in your body carries the story of these moments. You are the culmination of a chain of events so intricate and interconnected it defies imagination. From the stars to the soil, from the oceans to the trees, everything that came before whispers of your place in this grand design.

This is your beginning, and it is the universe's story too. As we explore the quantum mind of God, we may find that this incredible journey, from stardust to sentience, was no accident. It was an invitation to see ourselves as part of the greater whole, written into the symphony of creation itself.

A PRELUDE TO WONDER

What if the universe were not a cold, mechanical expanse but a living, breathing symphony, a divine dance where every particle sings its note, and every life adds its melody? What if science and spirituality, often seen as opposites, were threads in the same tapestry, weaving a story of profound interconnectedness?

This book invites you to explore that possibility. Through the lens of quantum mechanics, we glimpse a world where observation shapes reality, where entanglement connects the seemingly separate, and where the timeless "now" holds infinite potential. Through the wisdom of spiritual traditions, we see reflections of a universal truth: that humanity is not separate from creation but an integral part of its unfolding.

At its heart, this book is a journey, a journey to understand the Quantum Mind of God. It is a call to align with the rhythms of life, to heal the web of connection that sustains us, and to step into our role as co-creators of a flourishing future.

This is not a book of easy answers or rigid doctrines. It is an invitation to wonder, to question, and to see the world with new eyes. It is a bridge between the scientific and the sacred, the known and the infinite, offering glimpses of a universe alive with meaning and possibility.

Whether you read every page or only the beginning, may you carry this thought with you: the kingdom of God is at hand, not in the distant heavens but here, within you, waiting to be realized. And as you walk forward, remember, you are not alone. You are part of a symphony, a thread in the tapestry, a reflection of the divine.

CHAPTER 1: THE QUANTUM BRIDGE OF WONDER

At the quantum level, alignment is not a passive state but an active process. Particles interact through forces like electromagnetism, gravity, and the strong and weak nuclear forces, seeking a balance that allows them to form stable structures. This process underpins the very existence of matter.

But the journey from potential to reality is not always smooth. Take the phenomenon of wave-particle duality: a single particle, like a photon or an electron, exists as a wave of potential, a field of probabilities, until observed. The moment it is observed, when it interacts with another particle or is measured by a conscious observer, it collapses into a definite state.

This collapse is not random. It follows the rules of alignment, shaped by the forces, energies, and relationships present in that moment. The universe, it seems, is not just a collection of particles but a symphony of interactions, each note finding its place in the harmony of creation.

At its core, quantum alignment reveals a profound truth: the universe seeks balance. Imbalance is not ignored; it becomes the catalyst for change. Forces adjust, particles realign, and systems evolve, all to preserve the resonance that sustains creation.

But what happens when a part of the system deliberately steps out of alignment?

This is the paradox of humanity. Unlike particles, atoms, or even ecosystems, we possess the unique ability to observe the system not only from within but as though from the outside. This perspective has given us incredible insight, the ability to question, analyze, and shape our environment. It has also allowed us to make choices that disrupt the natural order.

From the first conscious decisions that marked humanity's separation, from the garden, from nature, from one another, we began to explore the boundaries of our own will. In seeking to understand the system, we stepped outside its flow, observing it not as participants but as detached spectators. This shift was both our greatest gift and our greatest challenge.

Yet, the universe does not leave imbalance unchecked. Just as particles correct their trajectories or ecosystems adapt to disruptions, creation unfolds in a way that brings even the most chaotic elements back toward harmony. The imbalance that does not serve the greater system cannot remain indefinitely.

Humanity's choices have created waves of imbalance, rippling outward through ecosystems, relationships, and even our own consciousness. But the universe responds, not with destruction, but with the gentle (and sometimes not-so-gentle) push toward correction. These corrections are not punishment; they are opportunities to realign, to remember our place in the resonance of creation.

As humanity stepped outside the natural flow, it was not without consequence. Our desire to observe, analyze, and manipulate the system brought incredible advances, but it also created dissonance. The harmony that had sustained life for eons now faced disruption, not from the universe itself, but from the choices of its most conscious participants.

This dissonance can be seen everywhere: in the degradation of ecosystems, the divisions between cultures, and the inner struggles of the human mind. Yet, even during this imbalance, we are not without guidance. The universe continues to speak, offering reflections of its natural order in the form of patterns, cycles, and resonances that echo through all things.

Modern science has become one of the tools through which we are beginning to hear this message again. The quantum revolution has opened a window into the very fabric of existence, revealing truths that ancient traditions have long intuited: everything is connected, nothing exists in isolation, and balance is not static but dynamic, a continuous process of realignment.

Quantum mechanics teaches us that even the tiniest particles are in constant communication, entangled across vast distances, responding instantaneously to changes in their counterparts. This phenomenon, once dismissed as impossible, now challenges our understanding of reality itself. How can two particles, separated by light-years, remain connected?

Perhaps the answer lies not in the particles themselves but in the system that binds them. The universe, it seems, is not a collection of separate parts but an integrated whole, a field of infinite possibilities where everything is interwoven.

This realization marks a turning point; a revolution not just in science but in how we understand our place within creation. The tools we once used to separate ourselves from the system are now revealing its profound unity.

The quantum revolution began with questions that seemed impossible to answer: How does light behave? What is the nature of matter at its smallest scale? These questions shattered classical notions of reality, revealing a world where particles could exist as waves, where probabilities replaced certainties, and where observation itself influenced the outcome.

One of the most striking discoveries was quantum entanglement, the phenomenon where two particles, once connected, remain inexplicably linked, no matter how far apart they are. When one particle changes, the other responds instantaneously, as though distance itself were an illusion. Albert Einstein famously called this "spooky action at a distance," yet today it is a proven and measurable reality.

Entanglement challenges everything we thought we knew about the universe. It suggests that separation is not fundamental, but rather an illusion created by our limited perspective. At the quantum level, connection is the rule, not the exception.

Entanglement is often described as a persistent connection between particles, but what if it's something even more dynamic? Recent discussions and experiments suggest that entanglement may not represent a continuous state but rather a moment of data transfer, a harmonization, if you will, between particles across spacetime.

In this view, the particles themselves are not holding the information. Instead, they are accessing something beyond themselves, a deeper memory or field of communication that binds the universe together. When one particle's state changes and the other responds instantaneously, it hints at a system where information does not travel but already exists in an interconnected whole.

This perspective challenges the idea that particles are the fundamental building blocks of reality. Instead, they may be the points where a much larger network of information becomes visible. The particles are like ripples on the surface of a vast ocean; the ocean itself, the underlying field, carries the memory, the data, and the connection.

If entanglement is a moment of communication rather than a static state, it suggests that the universe is not merely a collection of interacting parts but a living network, constantly exchanging information. This aligns with the concept of quantum fields as the true fabric of reality, but it also raises profound questions:

Where is this information stored?

Is the "memory" of the system encoded within the quantum fields themselves, or is there an even deeper layer of reality that we have yet to discover?

These questions hint at a universe that is both deeply connected and deeply intelligent. Entangled particles do not "decide" to align on their own; they are part of a larger, unseen system that orchestrates these interactions.

The idea of a hidden memory or field communicating through moments of entanglement also invites reflection on the nature of consciousness. If every particle is part of this interconnected network, then what role does observation play? Could consciousness itself be an extension of this universal memory, a way for the system to observe, reflect, and refine its own resonance?

This brings us full circle to the human experience. Our choices, actions, and thoughts are not isolated events. They ripple through the quantum fields, participating in the data exchange that sustains the universe. And just as particles align in moments of entanglement, we have the opportunity to align, not with chaos, but with the harmony of creation.

This revelation has profound implications, not just for science but for our understanding of existence. If the universe is inherently connected, then every action, every choice, ripples outward, influencing the whole. In this light, alignment takes on new significance, not just as a principle of matter and energy, but as a guiding force for conscious beings.

To the quantum explorers, these discoveries may seem like the beginning of a new frontier. But they also echo ancient wisdom. Spiritual traditions have long spoken of the interconnectedness of all things, of the unity that lies beneath apparent diversity. Now, for the first time, science is catching up, providing a language to describe what mystics have intuited for millennia.

As we navigate this quantum revolution, we are faced with both wonder and responsibility. The universe has handed us the tools to understand its workings, to glimpse the resonance of creation unfolding. But with this understanding comes a choice: will we use this knowledge to realign with the harmony of the system, or will we continue to create dissonance?

Today, a growing number of scientists and philosophers challenge the divide between science and spirit, offering glimpses of a deeper unity beneath the surface of existence:

David Bohm, a physicist and visionary, described the universe as an "undivided whole." His theory of the implicate order proposed that everything emerges from a deeper, unseen reality, a concept that resonates profoundly with spiritual notions of oneness.

Roger Penrose and Stuart Hameroff explored quantum consciousness, proposing that the brain's microtubules might process quantum information. Their work hints at a connection between human consciousness and the quantum field.

Fritjof Capra, in *The Tao of Physics*, drew striking parallels between quantum mechanics and Eastern mysticism, emphasizing that both see the universe as a dynamic, interconnected whole in constant flux.

These thinkers do not offer definitive answers. Instead, they invite us to consider a universe where science and spirituality coexist, not as opposing forces, but as complementary ways of knowing.

At its core, the quantum revolution reminds us that reality is not fixed but fluid, shaped by interaction and observation. This mirrors the teachings of spiritual traditions, which describe a hidden order beneath the surface of existence, a resonance that we sense but cannot always see.

To live in a universe where science and spirit meet is to embrace mystery. It is to see ourselves not as separate from creation, but as participants within it, shaping and being shaped by the unfolding resonance of infinite possibilities.

As we stand on the threshold of this understanding, we are left with questions that refuse to be contained by simple answers. What does it mean to live in harmony with the system? How might these insights transform the way we see ourselves, our relationships, and our place in the cosmos? These are not conclusions but reflections, threads to carry forward as we continue this journey.

There is a mystery at the heart of existence, a paradox that science, philosophy, and spirituality have all sought to unravel. It is the question of how something so vast, so intricate, and so interconnected could come to be. It is the question of why we are here and what role we play in a universe that is, at once, infinite and intimate.

The quantum revolution has given us new tools to explore this mystery, revealing the interconnectedness of all things. From the atoms in our bodies, forged in the hearts of stars, to the quantum fields that ripple through the fabric of reality, every aspect of existence tells a story of connection and creation.

And yet, here you are one observer among billions, connected to everything, yet unique in your perspective. You are the universe made conscious, a way for creation to reflect upon itself.

This consciousness does not belong to you alone. It pulses in the rhythm of waves, the hum of electrons, the silent growth of trees. Reality itself participates in the act of observation, shaping and being shaped by countless perspectives across the cosmos.

Take a moment to consider: What does it mean to be part of something so vast that it defies understanding, yet so intimately bound to your being that every breath affirms its truth?

We do not claim to have answers only fragments of insight scattered across disciplines and centuries. What lies ahead is not a solution but an exploration of the quantum mind of God and the consciousness that binds all things. If you leave this chapter with more questions than answers, then perhaps you are exactly where you need to be. Breathe. Pause. And ask yourself: What is observing you?

CHAPTER 2: THE FABRIC OF REALITY

At first glance, the building blocks of existence seem simple. Particles form atoms, atoms form molecules, and molecules give rise to the DNA that shapes life as we know it. Yet, beneath this apparent simplicity lies a profound mystery, a fabric of reality woven from particles and waves, energy and information.

This is the story of how the universe organizes itself, turning chaos into complexity, potential into form. It begins at the smallest scale, with particles that are not particles at all but waves of possibility, vibrating within quantum fields that span the cosmos. These fields, invisible and all-encompassing, hold the potential for everything we see, feel, and are.

Every particle you encounter, whether an electron orbiting an atom or a photon of light streaming through space, is an expression of quantum fields. These particles are not solid objects but excitations, ripples that rise and fall like notes in a cosmic symphony.

Take, for example, the carbon atom. Without carbon, life as we know it could not exist. And yet, carbon itself is born from a delicate balance of forces inside stars, an interplay of gravity, heat, and quantum mechanics that forges heavier elements from simpler ones. But particles are only part of the story. They are the letters of the universe's language, and like letters, their meaning comes from how they combine.

Particles may be the letters, but waves are the rhythm, the unseen currents that guide and connect them. At its core, the universe is not made of matter but of movement. Everything, every atom, every molecule, every strand of DNA, exists in constant vibration, resonating with the energies around it.

These vibrations are not random. They follow patterns, frequencies, and harmonies that create structure and order. DNA, the blueprint of life, is a perfect example. Its double helix is not just a chemical structure but a symphony of aligned forces, where the vibrations of particles and waves create the stability needed for life to emerge and thrive.

If particles and waves are the language and rhythm of the universe, then DNA is its story, a narrative written in a code of four letters: A, T, C, and G. This code carries the memory of life, passed down through generations, tracing its lineage back to the first single-celled organisms.

Yet DNA is more than a repository of genetic information. Recent discoveries suggest that it may also act as a quantum antenna, receiving and transmitting signals from the environment. Just as particles and waves interact to create resonance, DNA may interact with quantum fields, aligning itself with the forces that shape life.

This raises profound questions: Could DNA be a bridge between the physical and the quantum, a way for the universe to encode and preserve the resonance of creation? Is it possible that the double helix, with its elegant simplicity, holds not just the blueprint for life but a deeper connection to the fabric of reality itself?

As we journey through the fabric of reality, we begin to see how deeply connected everything is. Particles form waves, waves create patterns, and patterns give rise to the complexity of life. Yet, at every level, the same principles apply, alignment, resonance, and the interplay of forces seeking balance.

In understanding this, we are not just uncovering the mechanics of the universe, we are glimpsing its meaning. The same forces that shape galaxies also shape us, and the same patterns that guide the quantum world echo in the spirals of DNA.

Imagine standing on a shoreline, watching waves crash against the sand. Each wave carries the memory of its journey, ripples stirred by distant winds, currents shaped by the pull of the moon. As the wave crests and recedes, it leaves behind patterns etched in the sand, a fleeting record of its passage.

Now, look beyond the shoreline. Consider the waves of light traveling from distant stars, waves of sound echoing across a canyon, or the waves of energy rippling through the quantum fields. Each wave carries a trace of its past, leaving behind an imprint that shapes what comes next.

If the present moment is the crest of the wave, then what is history? It is not gone, nor is it something we can touch. Yet it lingers, etched into the very fabric of the universe, shaping the reality we now inhabit.

Every wave leaves a trace pattern in the sand, a ripple in the water, a memory in the mind. The same is true for the waves of energy and matter that make up the cosmos. The past does not vanish; it leaves its imprint in the arrangement of particles, the alignment of forces, and the resonances that echo through time.

Consider the light from a distant star. That light began its journey billions of years ago, crossing the vastness of space to reach us. When we see it, we are not observing the star as it is now but as it was long ago. In this sense, the light is a trace, a message from the past, carried forward by the wave of the present.

This idea extends beyond the stars. The Earth itself carries the memory of its formation, encoded in its layers of rock and sediment. Each stratum tells a story of ancient seas, shifting continents, and life emerging from the depths.

Even within us, the past is preserved. The iron in your blood was forged in the heart of a dying star; the oxygen you breathe was produced by ancient cyanobacteria that transformed the planet's atmosphere. Your very existence is a testament to the waves of history, their traces woven into the present moment.

Yet these traces are not static. They are dynamic, constantly influencing the present and shaping the future. DNA, for example, carries the memory of life across generations, but it is not merely a passive record. It adapts, mutates, and evolves, responding to the environment and creating new possibilities.

In the quantum world, the traces of the past take on an even stranger form. Particles leave "shadows" in the fields they traverse, influencing the paths of other particles. These interactions create patterns that ripple outward, like the wake of a boat on a calm lake.

If history is the wake, then the present is the crest of the wave, constantly unfolding from the past while propelling us into the future. This is the resonance of creation, a dynamic interplay of memory, motion, and manifestation.

In this view, the past is not something distant or abstract. It is alive within us, within the particles and waves that form the fabric of reality. To understand the present, we must recognize the traces of the past, the resonances that continue to shape the unfolding wave of creation.

What does it mean to live in a universe where the past is never truly gone, where every moment carries the imprint of what came before? How might this understanding change the way we see ourselves, our history, and our role in the great unfolding of reality?

If the universe carries the traces of its history in particles, waves, and fields, then memory is humanity's way of leaving a mark on the fabric of existence. Memory is more than a biological function; it is the human trace, a record of experiences, choices, and moments that ripple outward, shaping not only our individual lives but the collective reality we inhabit.

Consider the neural networks in your brain. Each memory is a pattern, a specific alignment of neurons firing together. But memory is not static. It is alive, reshaped every time you recall it, like a wave interacting with the shore and carving new patterns in the sand. In this way, memory is both a reflection of the past and a participant in the present, influencing the choices and actions that shape the future.

But memory extends beyond the individual. Through stories, art, music, and culture, humanity preserves its collective memory, a shared resonance that binds us together across time. These traces carved in stone, written on paper, or stored in digital code are the waves we send into the future, ensuring that our experiences and insights are not lost but carried forward.

What makes memory extraordinary is how deeply it resonates with the universe itself. Just as particles leave traces in quantum fields and light carries the memory of distant stars, human memory echoes through time, influencing reality in ways both subtle and profound.

In this sense, memory is not merely a human phenomenon. It is a universal principle, a way for the past to linger, to inform, and to shape the present wave of existence. Whether encoded in DNA, etched into rock, or stored in the synapses of a brain, memory is the thread that connects what was with what is and what could be.

What does it mean to be a species capable of memory, of tracing the past and projecting it into the future? How might we use this ability not just to preserve but to align, to ensure that the traces we leave behind contribute to the resonance of creation unfolding?

These questions bring us closer to the heart of our exploration. For in the interplay of particles, waves, DNA, and memory, we find not only the mechanics of reality but its meaning, a tapestry where each thread matters, where every trace carries forward the story of existence.

From the moment humans learned to communicate, memory became more than an individual experience, it became collective. Stories told around ancient fires became oral histories, passed from generation to generation, evolving but never lost. Later, these stories found permanence in cave paintings, stone carvings, and the written word, creating a tapestry of memory that spans millennia.

Consider the Rosetta Stone, a single artifact that preserved the key to understanding ancient Egyptian hieroglyphs. Without it, an entire civilization's language and stories might have remained locked in silence. Or think of the Dead Sea Scrolls, fragile yet enduring, carrying echoes of spiritual thought from thousands of years ago into the present day.

These artifacts are more than relics; they are waves of memory, traces left by those who came before us. They remind us that memory is not just about preserving knowledge but about creating connection between past and present, between people and the cosmos.

But memory is not confined to human creation. The Earth itself holds memories in its strata, fossils, and magnetic fields. Tree rings tell the story of past seasons; glaciers hold the chemical signatures of ancient atmospheres. Even the cosmic background radiation, a faint hum that pervades the universe, is a memory of the Big Bang, a trace of the moment creation began.

In this way, memory is woven into the fabric of existence itself. Whether carved into rock, written in DNA, or observed in the heavens, memory is the universe's way of preserving its unfolding story.

If memory is the thread that connects what was with what is and what could be, then what might this mean for humanity's role in the cosmos? Our memories, personal and collective, do not exist in isolation. They are part of a greater resonance, one that extends beyond Earth and into the stars.

The Voyager spacecraft, launched decades ago, carries golden records etched with sounds and images of life on Earth, memories sent into the vastness of space, a message for any being that might one day encounter it. These records are more than scientific experiments; they are humanity's declaration that we exist, that we remember, and that we seek to connect.

CHAPTER 3: THE QUANTUM BRIDGE

To truly understand our place in the universe, we must first embrace the paradox of knowing and unknowing. Humanity, the universe, and consciousness itself are connected by threads of mystery, a quantum tapestry that invites us to explore not just what we are, but why we are.

The quantum bridge spans the divide between science and spirituality, two worlds often seen as incompatible. Yet, as we begin to examine the universe's building blocks, the particles, waves, and fields that shape reality, it becomes clear that these two realms are not opposites. Instead, they are reflections, each offering insights into the other. Science explains the mechanisms; spirituality gives them meaning.

The journey across this bridge starts with wonder. Consider light, for example, a duality of particle and wave, a paradox that defies our traditional understanding of matter. Light is both a messenger and a mystery, carrying the story of stars across the void of space to illuminate our world. In this, light becomes a symbol of connection, of the unseen forces that bind us to the cosmos and each other.

The first time you truly pause to stare at the stars, really *stare*, something shifts inside you. There's a quiet realization that the light dancing above has traveled millions of years to reach you. The photons streaming into your eyes might have begun their journey when dinosaurs roamed the Earth, carrying a message of a past so ancient it feels unreal.

That light connects you to a story far greater than your own. In that moment, you might feel small, a fleeting speck in the vastness of the cosmos. And yet, there's an undeniable sense of belonging. You are not apart from this universe; you are *part* of it. The iron in your blood, the oxygen in your lungs, the thoughts swirling in your mind, all came from the same source as those stars.

This is the quantum bridge in its simplest form: the realization that we are not observers of the universe but participants within it. Our thoughts, our actions, and even our questions ripple out, folding into the infinite resonance of creation unfolding.

This bridge is not built of stone or steel but of questions, questions that humanity has carried for millennia. What is the universe? Why does it exist? And how do we, seemingly fragile beings, fit into its infinite expanse?

Science has begun to answer some of these questions, revealing a world far stranger and more wondrous than we could have imagined. The universe, at its core, is not solid or static. It is dynamic, a vast ocean of energy, particles, and waves in constant motion. Quantum physics has shown us that even the smallest particles, quarks, electrons, photons, exist in a state of potential, a dance of possibilities until observed.

But what collapses this potential into reality? This is where the bridge invites us to step into the unknown. Observation is not merely seeing; it is participation. The act of observing transforms the universe, shaping it into something tangible. In this way, consciousness becomes not just a passive witness but an active creator.

And yet, this realization is not new. Long before the language of quarks and wave functions, ancient traditions spoke of a divine force, a creative breath that brings order to chaos. The Taoists called it the *Way*. Hindus described it as *Brahman*, the essence of all existence. And the Bible begins with a poetic resonance: *"And God said, 'Let there be light.'"*

Across time and cultures, we have seen the same patterns, the same truths, expressed in different ways. Science and spirituality, once divided by misunderstanding, are converging at this bridge. They do not contradict each other; they complement, illuminate, and expand upon one another. Together, they guide us toward the same ultimate truth: that we are part of an unfolding creation, a resonance of infinite possibilities.

Not everything happens at once. The infinite possibilities of the universe do not unfold in chaos but in an ordered, purposeful sequence. Why? Because creation, at its core, is a process, a resonance shaped by alignment. It is through alignment that the universe transitions from potential to reality, from the vast field of "what could be" to the tangible beauty of "what is."

At every level, alignment acts as the unseen architect of existence. From the smallest quantum particles to the largest galaxies, this process weaves chaos into order, possibility into form. To understand how creation unfolds, we must first explore this principle of alignment and the resonance it creates.

Alignment is the foundation of creation. At the quantum level, particles do not exist in isolation; they interact, entangle, and align, forming the building blocks of matter and energy. This is how the universe maintains its structure, a constant dance of forces seeking equilibrium.

Consider the hydrogen atom, the simplest and most abundant element in the universe. Its very existence depends on the balance of opposing forces: the positive charge of its nucleus and the negative charge of its single electron. Without this delicate alignment, hydrogen could not exist. Without hydrogen, there would be no stars, no water, no life.

This principle of alignment extends beyond the microscopic. Stars themselves are born from the gravitational alignment of dust and gas, collapsing into luminous spheres of energy. Galaxies form through the alignment of billions of stars, orbiting in vast spirals. Even at the scale of life on Earth, ecosystems thrive when organisms align in harmony, each playing its role in a greater web.

And yet, alignment is not always a given. It requires a balancing of forces, a movement toward harmony that often involves struggle, adjustment, and change. When alignment is achieved, the result is resonance, an amplification of energy, order, and potential.

What is it that observes both the dream and the waking world? This question leads us into the heart of the mystery, a mystery not only of perception but of participation. In both realms, consciousness plays an active role, shaping the experience even as it observes it.

Think of the moments just before waking, when the dream's vividness begins to fade. There is a quiet awareness that observes the shift, standing at the threshold between two realities. This awareness does not belong to the dream, nor is it entirely rooted in the waking world. It is something deeper, a presence that remains constant even as the scenes and sensations change.

This presence, this observer, is the thread that ties all experience together. It is what allows us to recognize a dream as a dream, to reflect on a memory, to imagine a future. And yet, for all its constancy, consciousness remains a mystery. We cannot touch it, measure it, or see it directly. It is both the most familiar and the most elusive aspect of existence.

At its core, consciousness is not just something we have; it is something we are. It is the essence of awareness, the silent witness to everything we experience. And yet, it does more than observe. Consciousness participates in the unfolding of reality, shaping it through perception, thought, and intention.

Consider this: in the quantum world, observation collapses potential into form. A particle exists as a wave of probabilities until it is measured, until it is observed. In that moment, the wave collapses, and the particle takes on a definite state.

Is consciousness the ultimate observer, the force that transforms possibility into reality? Or is it itself shaped by the universe, a reflection of the patterns and principles that govern existence?

These questions reveal not answers but possibilities, inviting us to explore consciousness not as a problem to be solved but as a mystery to be lived.

Just as consciousness navigates the realms of dreaming and waking, it also bridges the material and the metaphysical, the known and the unknown. Dreams show us that reality is not fixed but fluid, shaped by the mind's ability to imagine, create, and transform.

The universe, too, appears to be fluid, a dance of particles and waves, probabilities and patterns. At its smallest scales, reality is not solid but dynamic, a symphony of vibrations that align to create form. Consciousness, it seems, is part of this symphony, a thread woven into the fabric of existence.

In this view, the waking world and the dream are not separate. They are reflections of the same underlying reality, glimpses of a deeper truth that transcends both. Consciousness is the bridge, the observer and participant that ties it all together.

As we step further into this mystery, we are reminded that the greatest questions do not demand answers; they invite reflection. What is the nature of the mind that dreams, the awareness that wakes, the consciousness that observes both?

To explore consciousness is to explore ourselves, to see not only the threads of our own experience but the patterns that connect us to the universe. It is to recognize that we are not separate from the mystery, we are part of it, woven into its unfolding story.

Take a moment. Close your eyes. Imagine the moment before waking, the space where dream and reality meet. Feel the presence of the observer, the constant awareness that stands at the threshold.

Now open your eyes and ask: Who is watching?

What if the universe itself is a dreaming mind, a fractalized exploration of possibilities, living out its own infinite imagination? And what if God is the waking mind, the constant awareness that observes and participates in the dream?

In this view, the universe is not separate from its creator but an extension of that waking awareness, a dream where every element is imbued with the consciousness of the dreamer. Just as our waking mind holds the dreaming mind within it, God holds the universe, observing it, shaping it, and experiencing it simultaneously.

This analogy carries profound implications. If God is the waking mind and the universe is the dream, then we are not isolated observers. We are threads within the dream, fragments of the divine imagination exploring the vast tapestry of creation. Our consciousness is not apart from God's but a reflection of it, a smaller flame drawn from the greater fire.

Within this dream, there is a still, small voice, a whisper of the waking mind calling out to the dreaming mind. It is the voice of truth, often drowned out by the noise of bad choices, illusions, and distractions. Yet it persists, guiding us back toward alignment, reminding us of the unity that underlies the apparent multiplicity.

When we quiet the noise, when we listen, we find that this voice is not separate from us. It is our own consciousness speaking to us, the dream remembering the dreamer, the part reconnecting with the whole.

The idea of the universe as a fractalized dream offers a way to understand the interplay between the infinite and the finite, the waking mind and the dreaming mind. Fractals are patterns that repeat at every scale, infinitely complex yet governed by simple principles.

In the same way, the universe reflects the waking mind in every part of the dream. From the spiral of galaxies to the spirals in our DNA, from the vastness of the cosmos to the intricacies of human thought, every aspect of creation mirrors the whole.

This fractal nature means that God is not just the dreamer but every part of the dream. The stars, the oceans, the trees, and each of us, all are expressions of the same divine consciousness, living out the possibilities of existence in infinite variation.

If this is true, then our role within the dream takes on new significance. We are not merely participants; we are co-creators, shaping dreams through our choices, actions, and thoughts. The still, small voice is a reminder that we have the power to align with the waking mind, to live in harmony with the resonance of creation.

In listening to this voice, we begin to see the dream for what it is, not an illusion to escape but a masterpiece to be explored, understood, and cherished. We are not separate from the divine; we are threads in the same tapestry, waves in the same ocean, dreams in the same mind.

Take a moment to sit with this thought: What if you are both the dreamer and the dreamed? What if the universe itself is a fractalized expression of consciousness, exploring its infinite possibilities through you?

And if that is true, what might it mean to listen to the still, small voice, to recognize the waking mind within the dream?

The mystery of consciousness is not just about understanding the mind. It is about understanding the dream, the dreamer, and the shared resonance that binds them together.

The role of consciousness as the observer has long intrigued scientists, philosophers, and spiritual seekers alike. In the quantum world, observation is not a passive act, it is a creative force, collapsing waves of potential into tangible reality. This interplay between possibility and form lies at the heart of existence, and consciousness is its catalyst.

Imagine standing before a sea of possibilities, each one shimmering with potential. The act of observation is like casting a net into this sea, drawing forth a single outcome from the infinite.

But here's the mystery: the observer does not stand apart from the sea. They are part of it, shaped by the waves even as they shape the catch.

This dynamic relationship between observer and observed, possibility and form reveal a profound truth: consciousness is not separate from reality. It is intertwined with it, a thread that binds the universe together.

But consciousness does not exist in isolation. It is not confined to individual minds or single points of observation. Like particles entangled across vast distances, consciousness seems to operate as a shared field, resonating across individuals, species, and even the cosmos itself.

Consider the phenomenon of collective behavior in nature. Flocks of birds move like one, schools of fish turn in perfect unison, and forests communicate through vast networks of fungi. These are not random acts of coordination but evidence of a shared awareness, a field of connection that transcends the individual.

Humanity, too, participates in this shared field. When a group of people come together with a shared intention, whether to solve a problem, create art, or simply connect, their consciousnesses seem to align, amplifying their collective energy. This phenomenon, often described as being "in sync," is not just a metaphor. It reflects the deeper truth that we are not separate beings but interconnected nodes in the vast network of existence.

This shared field of consciousness offers a way to bridge the divide between matter and spirit, the tangible and the intangible. It suggests that the universe is not a collection of separate parts but a unified whole, where every particle, every wave, and every thought resonates with the others.

In this view, consciousness becomes the bridge between the material and the metaphysical. It is the thread that ties together the seen and the unseen, the observer and the observed, the dream and the waking mind.

This realization invites us to reflect on the nature of reality itself. If consciousness is not confined to the individual but shared across the cosmos, then what does it mean to be conscious? How might this awareness shape the way we see ourselves, our relationships, and our place in the universe?

Take a moment to consider the ripple effects of your thoughts, actions, and intentions. What you observe, you shape. What you focus on, you amplify. In this shared field of consciousness, every choice matters, not just for you but for the whole.

Consciousness is not just the observer. It is the participant, the creator, the thread that weaves together the infinite possibilities of existence. To explore consciousness is to explore the universe itself, to see not just the patterns that shape reality but the awareness that breathes life into them.

At its core, consciousness is the foundation of experience, a presence that underlies everything we see, feel, and know. It is the awareness that watches the rise and fall of thoughts, the stillness beneath the chaos, the thread that connects the waking and the dreaming, the observer and the observed.

But foundational awareness is more than a philosophical concept. It is a lived reality, one that reveals itself when we quiet the noise and return to the essence of being. Think of the moment when you close your eyes and take a deep breath. Beneath the movement of thoughts, beyond the pull of sensations, there is something constant, a presence that is simply aware.

This awareness is not confined to the individual. It is not "yours" or "mine." It is universal, a shared foundation upon which all experience is built. Just as waves rise and fall on the surface of the ocean, our thoughts and perceptions are expressions of a deeper, boundless awareness.

To truly understand this foundational awareness, we must look beyond the surface. Imagine a stormy sea, its waves crashing and churning in every direction. At first glance, the chaos seems overwhelming, a restless motion with no pattern or purpose.

But dive beneath the surface, and a different reality emerges. In the depths of the ocean, there is stillness, a calm that is untouched by the turbulence above. This stillness is the foundation, the constant presence that holds the movement of the waves.

In the same way, consciousness is the stillness beneath the waves of thought, emotion, and sensation. It is not defined by the turbulence but holds it, allowing it to rise and fall without being disturbed. This stillness is always present, waiting for us to notice it, to return to the foundation from which all experience arises.

This foundational awareness is more than a refuge from the chaos of the mind. It is the ground of being itself, the essence from which all things arise and to which all things return. In this awareness, there is no separation, no boundary between self and other, subject and object, dream and reality. There is only presence, a shared resonance that permeates all existence. This is the awareness that connects us to one another, to the universe, and to the divine.

Take a moment to pause. Close your eyes and breathe. Feel the waves of thought and sensation moving through you. Beneath them, notice the stillness, the constant awareness that holds it all. This is your foundation.

This foundational awareness is not something to attain; it is something to remember. It is the essence of who you are, the thread that ties you to the infinite, the resonance of creation unfolding within you.

At its core, consciousness is the ability to sense and respond, a thread of awareness that runs through all of existence, from the simplest organisms to the vast complexity of human thought. This foundational awareness is not confined to humans or even animals. It begins in the smallest forms of life, hinting at a profound connection between all living things and their environment.

Even the smallest organisms display this foundational awareness. A bacterium moving toward nutrients or fungi sharing resources across a vast underground network are not just mechanical processes, they are expressions of a form of thought, a harmony with the environment that speaks of purpose and adaptation.

This awareness, though simple, is profound. It reveals a truth about consciousness: it is not static or confined to higher forms of life. It is dynamic, present even in the earliest threads of existence, and always in dialogue with the world around it.

As life evolves, so does consciousness. Awareness deepens, moving from simple responses to intentional engagement with the world. Perceptual awareness emerges, a way of perceiving the environment and acting with purpose.

Consider a bird navigating its migration path, guided by the Earth's magnetic field, or a fish swimming in perfect synchrony with its school. These are not random acts but intentional interactions, dialogues with the environment that reveal a deeper level of knowing.

Perception is more than survival; it is the mind's way of engaging with the world, of understanding and being understood. It is awareness reaching out, creating connections that sustain life and shape the story of existence.

Beyond perception lies the realm of emotion, where consciousness begins to experience relationships and connections on a deeper level. Emotional awareness fosters bonds, builds social structures, and creates a sense of belonging.

Think of dolphins playing within their pod, the joy of their shared interactions, or elephants comforting one another in times of grief, their gentle touches and low rumbles resonating with empathy. These moments remind us that consciousness is not just about the self, it is about the relationships that sustain and enrich us.

Emotion is the bridge between self and other, a shared resonance that amplifies awareness and creates meaning.

From these foundations, consciousness evolves into self-awareness, the ability to reflect not only on one's existence but on one's place within the greater whole. Self-awareness is the question that shapes humanity: "Who am I?" and, more profoundly, "What am I a part of?"

It is the human capacity to see beyond the immediate, to contemplate our role in history, our impact on the world, and our responsibility for the future. Self-awareness is not an endpoint; it is a doorway, beginning that invites us to explore the larger tapestry of existence.

At its peak, consciousness steps beyond individuality and merges into the universal. Transcendent awareness is the realization of unity, the sense of being part of something infinite and interconnected.

This awareness emerges in moments of stillness, the peace of meditation, the awe of stargazing, the profound sense of connection felt in nature. It is a reminder that at its highest, consciousness is not confined, it is shared, a resonance that binds all things together.

As we explore these layers of awareness, we begin to see that consciousness is not just about depth but about form. The ways we experience awareness, its modalities, shape how we perceive the world and interact with it. Whether sensing nutrients as a bacterium, comforting a friend in grief, or contemplating the cosmos, consciousness takes many forms, each one reflecting the infinite possibilities of the universe itself.

What does it mean to hold all these layers of consciousness within us; to be aware of ourselves and our world, to feel emotion, to perceive connection, and to glimpse the infinite?

Consciousness is not a single thread; it is a tapestry, woven from the many ways we sense, respond, and engage with existence. It is both individual and shared, finite and infinite, a reflection of the universe's boundless creativity.

As we move forward, we will explore how this foundational awareness shapes our reality, our relationships, and our place in the cosmos. For to understand consciousness is not just to understand ourselves, it is to understand the dream, the dreamer, and the waking mind that holds it all.

Imagine this: A simple experiment with light, a beam directed at a thin barrier containing two slits. On the other side of the barrier lies a screen, ready to record the light's behavior. When the light shines through the slits, something extraordinary happens.

If you block one slit, light behaves predictably, like tiny particles. It strikes the screen in a straight line, a clear path traced from the slit to the screen. But when both slits are open, the light does something strange: instead of two straight lines, it creates a pattern of alternating bright and dark bands. This is an interference pattern, the kind you'd expect from waves overlapping each other.

It seems that light isn't just a particle, it's a wave, too.

But the real twist comes when you add an observer. Scientists place a detector at one of the slits to see which path the light takes. Suddenly, the interference pattern disappears. The act of observing forces the light to "choose" a single slit, behaving like a particle again. It's as if light knows it's being watched.

This is the double-slit experiment, a foundational moment in quantum physics. And it raises a staggering question: Does reality itself depend on observation? In quantum terms, observation collapses the wave of possibilities into a definite outcome. Before the light is observed, it exists in a state of potential, a quantum wave encompassing all possibilities. The moment we observe it, the wave "collapses," and the light takes a specific path.

This isn't just a curiosity of light; it's a principle that applies to all quantum particles. At its core, the universe is a field of potential, and consciousness seems to play a role in shaping what becomes real.

This mirrors something profound about human consciousness. Our thoughts, intentions, and attention act like quantum observers in the world around us. When we focus on something, an idea, a goal, a relationship, it collapses from the wave of possibility into the tangible reality of action and experience.

This concept that observation shapes reality is not new. Across cultures and traditions, ancient wisdom has hinted at this truth:

In Hinduism, the universe is seen as Maya, a grand illusion shaped by consciousness. Brahman, the ultimate reality, becomes tangible only through observation and interaction.

In Buddhism, mindfulness teaches that attention shapes experience. "What you dwell upon, you become," echoes the idea that consciousness participates in the creation of reality.

In Western Mysticism, thinkers like Emanuel Swedenborg explored the idea that the spiritual and material worlds are connected, with human thought influencing outcomes in both realms.

Each tradition offers a lens to see what the double-slit experiment reveals: our role as conscious participants in the unfolding of reality.

Here's where the metaphor of plasma begins to whisper into the narrative. If quantum potential is the invisible source of possibilities, plasma is its visible counterpart, showing us how energy and fields interact to create structure and light. Just as plasma responds dynamically to electromagnetic fields, reality responds to the focused observation of consciousness.

Plasma becomes a symbol of the universe's creative dialogue, a dance between intention and manifestation. This concept will deepen as we explore entanglement and the nature of time in later chapters, but for now, it plants the seed: the universe is alive with interaction, and observation is part of the system.

If observation changes reality, how does your attention shape your life?

What waves of possibility are waiting for you to collapse them into the "now"?

If you take two particles and let them interact, something remarkable happens. No matter how far apart they are, whether inches or light-years, they remain connected. Measure one particle, and the other instantly "knows." This is quantum entanglement, what Einstein famously called "spooky action at a distance."

At first glance, this phenomenon defies everything we know about the universe. How can two particles, separated by vast distances, communicate faster than the speed of light? Yet countless experiments confirm that this connection is real. It's as if the universe operates on a level deeper than space and time, where separation is an illusion, and all things are fundamentally connected.

Quantum entanglement reveals a startling truth: the universe is not a collection of isolated objects but a web of relationships. At its most fundamental level, reality is a network of connections, where every particle, every wave, every field is interwoven with the whole.

This resonates deeply with ancient wisdom:

Indigenous traditions describe the world as a web of life, where every being is connected to the whole.

Taoism teaches the interconnected flow of Yin and Yang, where nothing exists in isolation.

Buddhist philosophy speaks of interdependence, the idea that all things arise and exist through their relationship to others.

These teachings align beautifully with what science is uncovering: that the universe is not built of separate parts but of inseparable connections.

Entanglement also challenges our understanding of space and time. If two entangled particles can "communicate" instantly, it suggests that at a deeper level, space and time might not be as rigid as they seem. Some scientists even propose that entanglement could be the foundation of spacetime itself, with the web of quantum connections weaving the very fabric of reality.

This opens profound questions:

Could consciousness itself be entangled with the universe, connected to all things at once?

Is the act of observation not just local but part of a universal network, where each observer influences the whole?

While entanglement operates at the quantum level, it offers a powerful metaphor for human experience. Just as entangled particles remain connected across vast distances, so too do our relationships, actions, and thoughts ripple through the web of existence.

Consider:

The way a kind word or a harmful act can echo through a community, affecting people we may never meet.

How the choices we make today shape the future, not just for ourselves but for generations to come.

The idea that we are never truly alone; our existence is part of something greater, bound by threads we cannot see.

Entanglement invites us to see the universe not as a machine but as a living, breathing network of connection. In this view, the God Mind is not a distant force but the very fabric of the web, present in every connection, every interaction, every thread that binds the cosmos together.

If we are entangled with the universe, then our role is not passive. We are participants in this divine network, shaping it with our thoughts, actions, and intentions. And just as entangled particles respond to each other, the universe responds to us, reflecting the energy we put into it.

As we begin to see ourselves as part of this web, we're left with questions that go beyond science:

What does it mean to live as though we are all entangled?

How might this change the way we treat each other, the planet, and ourselves?

If the threads of the universe are alive with connection, what kind of thread are you weaving?

As we delve deeper into the universe's web of connection, another profound question arises: how does this web interact with time? If entanglement suggests that particles are linked beyond the constraints of distance, could time itself be part of this universal connection?

In the same way that space seems to dissolve in the face of entanglement, time too begins to appear not as a rigid flow but as a dynamic, malleable dimension. To understand the nature of reality, we must explore how the eternal now shapes our perception of past and future.

What is time? To most of us, it feels like a river, flowing steadily forward, carrying us from the past into the future. But quantum physics suggests a reality far stranger: time might not flow at all. Instead, everything that has ever happened is happening, and what could happen exists in a state of potential, waiting for observation to bring it into focus.

In the quantum world, particles don't follow our familiar sense of cause and effect. Experiments like the delayed-choice quantum eraser show that choices we make in the present can influence how a particle behaved in the past. It's as if time itself bends to the act of observation.

This challenges our traditional understanding of time as linear. Instead, time in the quantum realm seems more like a canvas, where moments can be rearranged, connected, or even rewritten by interaction and intention.

If time is not a straight line, what is it? Many physicists and philosophers suggest that the "now" is all there truly is. The past exists as traces in the present, memories, records, fossils. The future exists as pure potential, shaped by the choices and intentions of the present moment.

This aligns with spiritual teachings across cultures:

In Buddhism, mindfulness centers on being fully present in the now, the only moment where reality unfolds.

In Christianity, Jesus says, "Do not worry about tomorrow, for tomorrow will worry about itself," emphasizing the sacredness of the present.

Indigenous wisdom often speaks of time as cyclical, where the past and future meet in the eternal now.

The intersection of these views suggests a profound truth: the "now" is not just a fleeting moment but the very point where potential becomes reality.

Modern physics proposes that time might be woven into the fabric of the universe, inseparable from space. This spacetime fabric stretches, bends, and even knots in the presence of energy and matter. Quantum entanglement hints that connections across this fabric might transcend our ordinary sense of time, binding moments and events together in ways we're only beginning to understand.

This view opens intriguing possibilities:

Could the "now" be the crest of a wave on the quantum fabric, where the potential of spacetime collapses into the observable present?

If the past is fixed and the future is malleable, how does intention shape what lies ahead?

For humanity, time is both a gift and a mystery:

Memory connects us to the past, allowing us to learn and grow.

Imagination reaches into the future, enabling us to dream and create.

But it is in the present, the now, where we act, observe, and participate in shaping reality.

Science shows us that our brains are wired to perceive time subjectively, stretching or compressing it based on attention and emotion. When we are deeply engaged or in a state of flow, time seems to vanish. This hints at a deeper truth: the more fully we live in the now, the more timeless we become.

If the universe is a web of entangled connections, time may be one of its threads, weaving moments together into a tapestry of meaning. The God Mind, in this view, could be the eternal observer, seeing all threads at once, past, present, and future, as part of a unified whole.

Yet, within this grand design, we are not passive observers. We are co-weavers, shaping the threads of time with our intentions and choices. Every action we take in the now ripples across the web, influencing not only the future but how the past is remembered and understood.

How might your life change if you saw the now as the only true moment of creation?

What traces of the past are you carrying into the present, and how are they shaping your choices?

If the future is a wave of potential, what intentions will you set to guide it toward harmony?

At this point, the pieces start to fit together: observation shapes reality, connections weave the universe into a cohesive whole, and time emerges as the eternal now, a canvas for creation. These are not just abstract ideas or scientific curiosities, they point to something deeply profound about our place in the universe.

We are not mere observers of the cosmos. The very act of observing connects us to it, making us participants in its unfolding. Just as quantum particles respond to observation, the universe responds to us. Every choice, every intention, every action we take sends ripples through the web of existence.

This means that:

We are deeply interconnected with one another and with the world. Our actions and choices resonate far beyond what we can see.

The now is our point of power. It is the moment where we translate potential into the reality of creation.

If we are interconnected with the universe, then it reflects back what we give to it. This idea resonates across traditions:

In Christianity, the principle of "reaping what you sow" suggests that the energy we put into the world returns to us.

In Hinduism and Buddhism, karma speaks to the cyclical nature of actions and their consequences.

Even in physics, the feedback loop of cause and effect shapes systems at every level.

This reflection invites us to consider: What are we creating? What kind of world are we observing into existence?

If the God Mind is the eternal observer, then humanity plays a unique role as co-creators within this divine network. Much like the interplay of fields guiding plasma into patterns of light and structure, our intentions shape reality:

Our consciousness influences the quantum field, just as electromagnetic fields guide plasma into visible forms.

Our actions and choices weave the fabric of reality, connecting us to past and future in a shared web of meaning.

This is not a call to perfection but to alignment. When we align our intentions with the universal flow, the outcomes are harmonious. When we act from disconnection or selfishness, the ripples create an imbalance not just for us but for the entire system.

Imagine the universe as a vast, living tapestry, woven from threads of connection, time, and intention. Every thought, every action, every observation is a stitch in this grand design. The God Mind, the ultimate weaver, invites us to take part in creation, adding our unique patterns to the whole.

But with this invitation comes responsibility. If our thoughts and actions are threads in the universal fabric, what kind of tapestry are we helping to create? Are we weaving patterns of love, unity, and harmony, or are we creating knots of fear, division, and imbalance?

As we conclude this chapter, we leave with these questions:

What is your role in the cosmic dance? Are you a passive observer, or are you consciously participating in the creation of reality?

How can you align your intentions with the greater flow of the universe?

If the present moment is your canvas, what will you choose to create?

The journey doesn't end here. In the next chapters, we'll dive deeper into the divine algorithm that underpins the universe, the intricate connections of life, and the infinite possibilities waiting to unfold. But for now, let's rest in the wonder of what we've glimpsed: a universe alive with interaction, shaped by consciousness, and brimming with potential.

CHAPTER 4: THE DIVINE ALGORITHM

The divine algorithm isn't just a system, it is alive with intention and connection. This brings us to a profound possibility: the God Mind, the source of all creation, might be experienced as a field of infinite energy and intelligence, permeating all existence.

If the God Mind can be imagined as a vast electromagnetic field of intention, then humanity might be seen as localized fields within it, connected but unique. The human heart, generating the strongest electromagnetic field in the body, may act as the bridge, resonating with this divine source.

What we often call the Holy Spirit, or the Spirit, might be understood as God Source Energy; the living connection between the infinite mind of God and the embodied human experience. Like an electric current passing through a circuit, this energy flows between us and the divine, aligning our will with the universal algorithm.

The universe behaves like a dynamic system, constantly evolving and recalculating through the interactions of its components. At every level, from subatomic particles to vast galactic structures, we see patterns of interaction that hint at a deeper organizing principle. These interactions reveal a system that is not static but alive, responsive, and adaptive.

Here's how these interactions manifest:

Quantum Rules: At the smallest scales, particles interact probabilistically, guided by quantum rules that govern their behavior. These rules allow for vast potentialities, which collapse into reality only when observed or measured.

Self-Organization: In nature, systems spontaneously form

patterns, from snowflakes to spiral galaxies. This phenomenon of self-organization suggests that the universe operates with an inherent tendency toward order and complexity, even amidst chaos.

Fractal Patterns: The same principles appear on every scale of the universe. Just as a fractal replicates its pattern infinitely, the interactions that shape an atom also influence ecosystems, planetary systems, and galaxies.

Dynamic Feedback Loops: The universe thrives on interaction. Stars create the elements that form planets, which in turn give rise to life that eventually looks back at the stars. This feedback loop is not just physical but participatory, with humanity itself becoming part of the universe's evolving narrative.

These dynamic interactions mirror the concept of an algorithm, a process that solves problems or generates outcomes through iterative steps. The universe operates much like an open algorithm:

Input: Forces, particles, and fields act as variables entering the system.

Processing: Interactions, guided by natural laws and quantum principles, continuously compute the relationships between these inputs.

Output: The observable universe, with all its patterns, structures, and emergent phenomena.

But unlike human-designed algorithms, which follow rigid rules, the universe's algorithm is inherently flexible. It adapts, evolves, and grows, always recalibrating based on new inputs.

The universe's interactions reveal a constant dance between chaos and order. In moments of chaos, energy disperses, and systems break apart. Yet, out of this chaos, new forms emerge, stars are born from collapsing gas clouds, ecosystems rebound after disruption, and life itself thrives on adaptation.

This duality echoes throughout the cosmos, suggesting that the universe is a system not of fixed rules but of dynamic balance.

This interplay of forces invites us to consider a deeper question: Is the universe purely mechanical, or is there intention behind its interactions? Across traditions, we find the belief that a divine intelligence, a Logos or guiding principle, weaves through this system. Whether seen as a cosmic algorithm or a sacred dance, the universe's interactions point to a profound truth: creation is not a one-time event but an ongoing process, alive with potential and purpose.

Across cultures and traditions, humanity has long sought to understand the underlying order of the universe. What science calls the laws of nature, many spiritual traditions have described as divine principles, forces that not only govern reality but imbue it with purpose. At the heart of these teachings is the belief in an ultimate organizing force, often referred to as the Logos, Tao, or Dharma.

In Christianity, the Logos is described as the divine Word, the principle through which all things were made. John's Gospel opens with profound simplicity:

"In the beginning was the Word, and the Word was with God, and the Word was God. Through Him all things were made; without Him, nothing was made that has been made." (John 1:1-3)

The Logos is not just a creative force but a sustaining one, upholding the cosmos and infusing it with meaning. It is both the algorithmic framework of the universe and the divine intention behind it, uniting structure with purpose.

In Hinduism, the concept of Dharma reflects a universal order, a cosmic law that governs all things. Dharma maintains harmony and balance, ensuring that the universe functions as an interconnected whole. It extends to moral and ethical principles, suggesting that human actions must align with this cosmic order to maintain equilibrium.

Similarly, Buddhism describes the workings of karma—a system of cause and effect that mirrors the algorithmic nature of the universe. Every action generates ripples, creating outcomes that reinforce the interconnectedness of all things.

In Taoism, the universe is governed by the Tao, the Way. The Tao is not a personified god but the fundamental principle of balance and flow. It is both the source and the path, the dynamic process through which the universe unfolds. The Tao reminds us that to align with the universe, one must move in harmony with its rhythms, adapting without resistance.

Despite cultural differences, these traditions share a profound insight: the universe is not chaotic or random. It is governed by principles of order, balance, and meaning. Whether described as the Logos, Dharma, or Tao, these concepts suggest a deep, unifying truth: the universe is both structured and intentional, a reflection of divine intelligence.

From a scientific perspective, this divine order mirrors the universe's dynamic algorithm:

The Logos aligns with the rules and constants of nature, providing the framework for creation.

Dharma and karma resemble the feedback loops within systems, where actions generate consequences that ripple outward.

The Tao reflects the adaptive flow of the universe, balancing chaos and order to sustain life and growth.

These parallels suggest that what science describes as natural laws may be the very principles that ancient traditions intuited as divine. The language differs, but the truth remains the same: the universe is alive with connection, pattern, and purpose.

If the God Mind is the ultimate source, then these spiritual principles are its expression. They are the visible patterns of an invisible intelligence, the tangible algorithms of a divine will. Through the Logos, Dharma, and Tao, the God Mind speaks, weaving creation into existence and inviting humanity to participate in its unfolding.

As we explore the dynamic system of the universe and the principles of divine order, a deeper question emerges: How are we, as conscious beings, connected to this system? What bridges the infinite mind of the universe, what we've called the God Mind, and the lived human experience? The answer may lie in a universal connection we often feel but cannot fully articulate, a force that flows through and binds all things: God Source Energy.

Imagine the God Mind as a vast field of infinite energy and intelligence, like an electromagnetic field encompassing all of existence. This field is not just abstract or distant; it is alive with intention, purpose, and love. It is the source from which everything emanates, sustaining the cosmic algorithm and imbuing it with meaning.

If the universe operates as a dynamic algorithm, then the God Mind could be thought of as its programmer, its energy source, and its guiding intelligence. Like an electric current activating a circuit, the God Mind flows through all creation, shaping and sustaining reality.

Within this grand field of divine energy, humanity occupies a unique role. Our physical bodies, much like the universe, are dynamic systems. At the center of this system is the heart, which generates the strongest electromagnetic field (EMF) in the body, far exceeding that of the brain.

The heart's EMF can be seen as the bridge between the human self and the divine field. It is through this energy that we connect, align, and resonate with the God Mind. This idea is echoed in spiritual traditions that describe the heart as the seat of the soul or the dwelling place of the Spirit.

What we often call the Holy Spirit, The Great Spirit, the breath of life, the divine presence within, may be understood as God Source Energy. This energy flows from the God Mind, connecting all things in a web of unity:

Transcendent: It is infinite and universal, the same energy that powers the stars and breathes life into galaxies.

Immanent: It resides within each of us, aligning our being with divine order when we allow it to flow freely.

In this view, Spirit is not a separate force; it is the living connection between the infinite and the finite, the universal and the personal. It is the energy that animates the cosmic algorithm and invites us to participate in its unfolding.

To align with the God Mind is to tune ourselves to the frequencies of love, compassion, and purpose, because these states reflect the natural rhythms of the universe. Love creates connection where there is separation. Compassion restores harmony where there is imbalance. Purpose gives direction to the flow of energy, aligning our actions with the greater good.

This is not a philosophical ideal, but a pattern observed across scales:

In the body, love and gratitude create coherence, aligning our systems for optimal function.

In relationships, compassion fosters unity and mutual aid, mirroring the interconnected web of existence.

In nature, balance and cooperation sustain life, reflecting the universal drive toward harmony.

Spiritual traditions have long understood what science now observes: love and compassion are not just moral virtues but resonances that align us with the fundamental order of the cosmos. When we embody these frequencies, we allow the divine energy of the God Mind to flow through us, making us co-creators of a reality that reflects its balance and beauty.

Studies of heart rate variability show that emotions like love, gratitude, and compassion create coherent heart rhythms. These rhythms foster states of harmony within the body, aligning our systems for optimal function.

If resonance creates harmony in our physical systems, it's plausible that the same principles apply to how we connect to the larger cosmic field.

Acts of love and compassion create connection and unity between individuals, communities, and even nations. This mirrors the universe's tendency toward interconnectedness, where relationships, not isolation, form the basis of reality.

The universe thrives on connection, and love is a force that unites rather than divides.

Spiritual Wisdom Across Traditions

Christianity: Jesus emphasizes love as the greatest commandment, stating, *"Love your neighbor as yourself."* This reflects alignment with divine will, where love is the guiding principle of creation.

Hinduism and Buddhism: Compassion (Karuna) is seen as a path to enlightenment, aligning the self with the universal flow.

Taoism: Harmony with the Tao is achieved by flowing with the rhythms of nature, which include empathy and care for all things.

Across traditions, love and compassion are repeatedly described as aligning us with the divine or universal truth.

In physics, systems seek balance. Love and compassion mirror this universal tendency by restoring harmony where there is conflict or imbalance.

Nature thrives on cooperation; plants exchange nutrients, animals form symbiotic relationships, ecosystems depend on balance. Love and compassion in human behavior reflect this same cooperative principle.

Love is not just an abstract virtue; it reflects the universe's innate drive toward unity and growth. This perspective is not just spiritual but deeply grounded in what we observe scientifically:

The measurable energy of the heart can influence both the body and the environment, suggesting a tangible connection between the individual and the collective field.

The universe operates as a participatory system, where observation and interaction shape outcomes. The Spirit, as God Source Energy, could be the force that binds these interactions into coherence.

Studies show that emotions like gratitude, compassion, and love create coherent heart rhythms, fostering a state of flow that aligns with health and well-being.

These insights reveal a profound truth: the universe is not disconnected or indifferent. It is alive with connection, meaning, and intention, and we are active participants in its unfolding.

To understand the God Source Energy is to recognize that we are not separate from the divine algorithm. We are its expressions, its participants, and its co-creators. Spirit flows through us as both a gift and a responsibility, calling us to live with intention, love, and alignment.

As we move through life, the question becomes: How will we resonate with this energy? How will we align our hearts with the divine field, becoming conduits for creation and harmony in the cosmic algorithm?

What does this say about the tension between free will and determinism that has shaped philosophy, science, and spirituality for centuries? Are we living out a script written by the laws of the universe, or do we have the power to choose our own path? Quantum mechanics reframes this debate, offering a fresh perspective that blends structure with freedom, chaos with intention.

In the classical physics of Newton and Einstein, the universe was seen as deterministic. Every effect had a calculable cause, creating a chain of events that unfolded predictably. The universe was a mechanical clock, its gears turning according to fixed laws, with no room for spontaneity or deviation.

Under this view, free will seemed like an illusion; every decision we make, every thought we have, was predetermined by prior states of the universe. Humanity appeared as passengers, not participants, in the cosmic machine.

Quantum mechanics shattered the deterministic view. At the quantum level, particles do not follow predictable paths. Instead, they exist in states of potential, governed by probabilities. Observation collapses these probabilities into a definite state, suggesting that interaction shapes reality itself.

Entangled particles influence each other instantaneously, blurring the boundaries of cause and effect. This quantum framework introduces a paradoxical truth: Determinism still exists at a broader level. The universe operates according to mathematical principles and natural laws.

But within this framework, there is freedom of possibility, where outcomes are not fixed until observed or acted upon.

Quantum mechanics suggests that free will is not about breaking the laws of the universe but about participating within them. Our actions, intentions, and observations act as inputs into the cosmic algorithm, influencing outcomes.

Free will is the ability to engage with the universe's framework, shaping how potential becomes reality.

This dynamic interplay mirrors spiritual teachings:

In Hinduism, karma reflects the law of cause and effect, where free will exists within the bounds of cosmic order.

In Christianity, humanity is granted free will, but alignment with divine will leads to harmony and fulfillment.

In Taoism, the Tao provides the structure of existence, while individuals are free to flow with or resist it.

Free will and determinism are not opposites but partners in the dance of creation. The universe provides the structure; quantum rules, physical laws, and divine principles. Humanity provides the choice; our intentions, actions, and interactions with the system.

This relationship is not rigid but dynamic, evolving as we interact with the world. Like musicians improvising within the structure of a melody, we are free to create, but the music flows best when we stay in tune with the greater harmony.

If the God Mind is the source of the cosmic algorithm, then free will is its most profound gift to humanity. It allows us to participate in creation, shaping reality through our actions.

We can align with the divine flow. We can choose harmony over dissonance. We can learn and grow, as choices ripple through the system, teaching us the consequences of imbalance and the beauty of alignment.

The tension between free will and determinism becomes a source of creative potential, where humanity is invited to co-author the unfolding of the universe.

Quantum mechanics reframes free will not as absolute independence but as a dynamic relationship with the cosmos. The universe is alive with possibility, and we are participants in its story. Every choice, every observation, every action contributes to the ever-evolving dance of creation.

The universe operates as a living algorithm; a system of dynamic interactions shaped by laws, feedback loops, and intentions. But unlike human-designed algorithms, which are rigid and predefined, the divine algorithm of the cosmos is open, adaptive, and participatory. It invites humanity into its unfolding, making us not mere observers but active participants in the creative process.

Like any system, the divine algorithm has essential components that govern its function: Energy, Intention, and Action. Every action we take, every thought we think, every feeling we hold acts as an input into the system. These inputs ripple through the quantum field, influencing outcomes at both personal and cosmic scales.

Just as a drop of water creates ripples across a pond, our choices create waves that intersect and shape the larger flow of reality. The universe "processes" these inputs through its inherent rules:

Quantum mechanics governs the probabilistic behavior of particles.

Physical laws guide the interactions of matter and energy.

Divine principles ensure balance and meaning.

This processing is not static or mechanical. It is dynamic, responding to the unique inputs of each moment.

The outputs of this algorithm are the realities we experience, both personal and collective.

Whether it's the growth of a tree, the alignment of a planetary system, or the unfolding of a human life, the algorithm's outputs reflect the interplay of universal laws and individual contributions.

The Role of Humanity

While the divine algorithm operates at every level of existence, humanity plays a unique role as conscious participants. Unlike other beings, we have the capacity to reflect, choose, and direct our energy intentionally. This makes our inputs into the algorithm particularly potent.

As participants, we shape the algorithm's outputs, influencing not only our own lives but the collective experience of the world. The feedback we receive, whether harmony or discord, teaches us how to align more closely with the algorithm's flow.

The divine algorithm thrives on harmony, where inputs align with the greater flow of the universe.

However, imbalance occurs when inputs, such as fear, greed, or disconnection, create dissonance in the system. This dissonance disrupts not only individual lives but also the collective equilibrium. Yet the algorithm is self-correcting: Just as ecosystems restore balance after disruption, the universe works to realign itself over time.

Humanity's role is to recognize these imbalances and act in ways that restore harmony, both within us and in the larger system. The divine algorithm reflects timeless spiritual truths:

Karma: Actions generate consequences that ripple through the universe, mirroring the algorithm's feedback loops.

Dharma: Aligning with the cosmic flow ensures balance and fulfillment.

Logos: The organizing principle of creation sustains and guides the algorithm's processing.

These parallels suggest that the universe's algorithm is not only functional but meaningful, designed to foster growth, connection, and purpose. The divine algorithm is not a closed system; it is alive with potential.

While the algorithm provides structure, it also allows for infinite variation and creativity. Every moment is a chance to contribute something new to the system.

The closer our inputs resonate with the universe's principles, love, compassion, and purpose, the more harmonious the outcomes.

This makes the algorithm not a machine but a conversation, where humanity's choices shape the unfolding of reality in collaboration with the God Mind.

In the grand cosmic algorithm, humanity plays a unique and irreplaceable role. While the universe is alive with dynamic interactions, from the smallest particle to the largest galaxy, humans contribute something extraordinary: the ability to reflect, choose, and intentionally shape reality. This capacity elevates us from participants to co-creators in the unfolding story of existence.

The divine algorithm operates regardless of our awareness, but human consciousness amplifies its creative potential. Through observation, intention, and action, we act as catalysts within the system. Like the observer effect in quantum mechanics, our awareness focuses and collapses potential into reality. What we give our attention to becomes part of the system's output.

Our thoughts and emotions shape the energy we contribute to the algorithm. Love, compassion, and purpose resonate with the system's harmony, while fear and disconnection create dissonance.

Choices translate potential into tangible outcomes, influencing not only our own lives but the collective flow of the universe.

Human consciousness is both a gift and a responsibility, calling us to engage with the algorithm in ways that align with its deeper rhythms. If the God Mind is the source of the algorithm, then the human heart is its bridge. As the center of our electromagnetic field and the seat of love, compassion, and intuition, the heart connects us to the greater field of divine energy. When we act from the heart, we align ourselves with the God Mind, tuning into the frequencies that sustain harmony and growth.

This is why love and compassion are not just virtues but tools for resonance:

Love creates unity where there is division.

Compassion restores balance where there is suffering.

Purpose channels energy into meaningful creation.

These are the frequencies that allow the divine energy to flow through us, making us not just observers but active participants in the cosmic dance. Humanity's role in the algorithm reflects a profound balance between freedom and structure.

The framework is provided by the universe's laws and principles, which guide the flow of energy and matter. Freedom comes from our ability to choose how we interact with this framework, adding our unique inputs to the system.

This interplay echoes spiritual teachings:

In Christianity, free will is a sacred gift, meant to align with God's will for love and justice.

In Hinduism and Buddhism, karma emphasizes that freedom exists within the bounds of cosmic order, teaching us to act with wisdom and care.

In Taoism, Tao provides a path, but walking it is a choice.

Our freedom is not limitless, but it is profound. It is the opportunity to bring intention, creativity, and purpose into the system, shaping outcomes in ways no other being can.

Every action we take sends ripples through the algorithm, influencing the collective experience of the world. Just as a single drop of water can create waves across a pond, our choices affect not only ourselves but also the larger system.

A kind word can inspire hope in someone who inspires change in a community. An act of generosity can set off a chain reaction of compassion and cooperation. Conversely, a moment of harm or selfishness can ripple outward, creating imbalance and suffering.

The ripple effect reminds us that our role in the algorithm is not isolated. We are part of an interconnected web, where every choice matters.

To fulfill our role in the algorithm, we are called to align our thoughts, intentions, and actions with the divine flow. This alignment is not about perfection but about participation.

The heart is our compass, guiding us toward love, compassion, and purpose. In every moment, we have the choice to contribute balance or dissonance to the system.

As Peter Marshall, the Scottish-American preacher and Chaplain of the U.S. Senate, prayed in 1947:
"May we think of freedom not as the right to do as we please, but as the

opportunity to do what is right."

Freedom is a chance to resonate with the divine. When we align with the God Mind, we become conduits for its energy, co-creators in the ongoing process of creation. This is the highest expression of our humanity, where we transcend self-interest and become part of something greater.

As we reflect on the divine algorithm of the universe, one truth stands out: we are not passive observers of a distant system. We are active participants, living within the algorithm and shaping it with our every thought, action, and intention. This profound relationship between humanity and the cosmos reveals both the beauty of our freedom and the weight of our responsibility.

The universe, with its intricate patterns and dynamic interactions, invites us into a dance of creation. This dance is not choreographed, it is improvised, evolving moment by moment as we engage with it. The God Mind provides the rhythm and the framework, but it is humanity that brings the steps, infusing the dance with creativity, purpose, and meaning.

To live within the algorithm is to recognize that every choice we make ripples outward, influencing the whole. Our intentions shape the energies we contribute, for better or worse.

By aligning with the God Mind, we participate in the unfolding of harmony, balance, and growth.

Freedom is the gift that allows us to engage with the algorithm consciously. But freedom is not the absence of constraint; it is the opportunity to choose alignment, to act with intention and love.

In this sense, freedom is both a gift and a calling: A gift, because it empowers us to shape our reality. A calling, because it asks us to contribute to the greater good.

This perspective transforms freedom from a self-centered pursuit into a sacred responsibility, one that connects us to the God Mind and the greater flow of existence. To live within the algorithm is to embrace both our freedom and our responsibility:

Freedom gives us the ability to choose, to create, and to grow. Responsibility calls us to act with love, compassion, and purpose, aligning our inputs with the harmony of the God Mind.

This alignment is not about perfection but about intention. It is a journey of resonance, where each moment offers a chance to reflect, adjust, and re-tune us to the divine flow.

The divine algorithm is not fixed, it is alive with possibility. Every now is a new beginning, a fresh canvas upon which we can paint our choices and intentions. This makes the universe not just a system but a conversation, one in which humanity and the God Mind co-create the future together.

We are part of something vast and wondrous, a tapestry of connection and meaning. And in this tapestry, our threads matter. By living with awareness, by aligning with the principles of love, compassion, and purpose, we weave a reality that reflects the beauty and balance of the God Mind itself.

How does your understanding of freedom shift when seen as an opportunity to do what is right?

What intentions will you bring into the algorithm, knowing that your choices ripple through the cosmos?

How can you align more fully with the divine flow, becoming a conscious co-creator in the dance of existence?

CHAPTER 5: ENTANGLEMENT

Imagine looking up at the night sky, at the countless stars scattered across the heavens. At first glance, they appear isolated, each one a solitary beacon in the vastness of space. But what if the stars are not separate at all? What if every point of light is connected to every other, not by visible threads, but by a deeper, unseen fabric, a web that unites the universe into a single, living system?

In the quantum realm, this idea is not just poetic, it is reality. Entanglement, one of the most mysterious and awe-inspiring phenomena in physics, reveals that particles can become so intertwined that they remain connected no matter how far apart they are. When one particle changes, the other responds instantly, as though distance itself doesn't exist.

Albert Einstein famously referred to entanglement as "spooky action at a distance," expressing his discomfort with the idea that particles could influence one another instantaneously across vast distances. This defied his deeply held belief in a deterministic universe governed by local interactions where effects are limited by the speed of light and occur only in immediate surroundings.

However, decades of experimental evidence have proven Einstein's skepticism unfounded. Groundbreaking research has shown that entangled particles behave as though they are part of the same unified system, regardless of the distance between them. When one particle's state is measured, the state of its entangled partner is instantly known, without any signal traveling between them.

This phenomenon has been repeatedly demonstrated in experiments testing Bell's theorem, which showed that no hidden variables or classical explanations could account for the observed outcomes. In 2022, researchers Clauser, Aspect, and Zeilinger received the Nobel Prize in Physics for their work confirming that entanglement is a fundamental feature of quantum reality, not an anomaly or quirk of theory.

Entanglement challenges two key classical assumptions about the nature of reality.

Locality: The idea that objects are only influenced by their immediate surroundings. Entanglement suggests that connections between particles can exist beyond spatial boundaries.

Separability: The belief that objects have independent properties, regardless of their relationship to other objects. Entangled particles, by contrast, share a state that only makes sense when viewed as part of a larger system.

This points to a deeper truth about the universe: at its most fundamental level, reality is not made up of isolated entities but of relationships. The properties of entangled particles emerge not from their individual characteristics but from their connection to one another.

One of the most profound implications of entanglement is its challenge to the classical view of space and time.

Entanglement demonstrates that the universe operates in ways that transcend spatial separation. This doesn't mean particles "communicate" faster than light but rather that their connection exists in a way that isn't bound by space. This suggests that the universe is fundamentally interconnected, where the boundaries we perceive between objects may be constructs of our limited perspective.

However, it is important to remain precise. The scientific reality of entanglement is confined to the quantum level, where it governs interactions between particles. Its implications for larger systems, such as human consciousness or the universe as a whole, remain speculative.

While the concept of entanglement is rooted in rigorous science, its principles invite us to think about the interconnectedness of the universe more broadly. Entanglement shows that relationships, not isolation, define the most basic elements of reality.

This echoes patterns we see in nature, where ecosystems, communities, and even human relationships thrive on connection and mutual influence.

Recognizing these parallels is not an attempt to conflate science with metaphor but to highlight how the principles of entanglement resonate with truths we observe on larger scales. This bridge between science and philosophy invites us to consider: if particles are so deeply connected, might all of existence be part of a greater web of connection?

This chapter will continue to explore entanglement both scientifically and philosophically. While maintaining the integrity of its quantum definition, we will carefully examine how its principles illuminate broader truths about the unity of existence, the nature of relationships, and humanity's place in the cosmos.

A Universe Woven Together

If entanglement reveals that particles are deeply connected, it also hints at a larger truth: the universe itself may be woven from threads of relationship. Every interaction, every connection, every ripple of energy contributes to a vast and intricate tapestry of existence.

This interconnected nature of reality challenges the classical view of the universe as a collection of separate objects. Instead, it suggests a relational universe, where the properties of individual parts emerge from their participation in a greater whole.

Entanglement demonstrates that particles can remain connected across vast distances, responding to each other as if separation does not exist. This phenomenon, known as nonlocality, suggests that space is not the ultimate barrier we perceive it to be. The fabric of the universe may be more tightly knit than we can imagine and interactions at the quantum level ripple outward, contributing to a universal field of connection.

While these connections are measurable at the quantum scale, they inspire deeper questions: Could the same principles apply at larger scales, from ecosystems to human consciousness? If so, the universe itself may function as a vast web, where every thread influences the pattern as a whole.

Picture the universe as a loom, where each thread represents a point of connection, and the God Mind is the master weaver. The design unfolds moment by moment, as the present meets the potential of the next thread. The tapestry of existence is not static; it evolves dynamically, shaped by every interaction and choice.

This imagery is not just poetic, it reflects the patterns we see in nature. In ecosystems, every species plays a role in maintaining balance, from pollinators to predators. In relationships, the bonds we form ripple outward, influencing others in ways we often cannot see. In communities, individual contributions weave together to create the strength and beauty of the collective. Each thread matters. Each connection contributes to the design.

The idea of a woven universe is not new; it resonates with spiritual teachings from across cultures:

Buddhism speaks of dependent origination, the idea that all things arise and exist through their relationships to others.

Indigenous wisdom often describes the world as a web of life, where every being is interdependent.

Christianity reflects this unity in verses like *"For just as each of us has one body with many members... so in Christ we, though many, form one body"* (Romans 12:4-5).

These teachings remind us that connection is not just a scientific principle but a spiritual truth, a reflection of the divine design.

The universe's interconnectedness makes it more than a collection of parts; it is a living system, vibrant and dynamic. Every thread of the web is in motion, contributing to the unfolding of the whole.

Actions create reactions, rippling through the system in cycles of cause and effect. Like a fractal, the same principles appear at every scale, from quantum particles to galactic structures.

This living system reflects the divine algorithm in action, where each thread is guided by principles of balance, harmony, and growth.

If the universe is a woven tapestry, then humanity is not merely an observer, we are threads in the loom, part of the design itself. This understanding calls us to recognize that every action contributes to the pattern, strengthens the threads that bind us to others, and to the divine principles of balance and harmony; adding beauty to the greater design.

The universe, like a loom, weaves the fabric of existence from threads of connection. Each thread is part of a design far greater than itself, shaped by the master weaver. To live within this tapestry is to embrace our role as participants in its creation, recognizing that every choice, every action, every connection contributes to the unfolding pattern.

If the universe is a woven web, then the God Mind is the weaver, the source of the threads, the loom that holds them, and the intelligence that guides the pattern. Through entanglement, we catch a glimpse of the God Mind's presence, not as a distant force but as a living energy that flows through every connection.

At the quantum level, entanglement demonstrates that separation is an illusion. Particles that are entangled act as though they are one, regardless of distance. This unity at the smallest scales mirrors the spiritual insight that the universe is not divided but whole. The God Mind could be seen as the field that sustains this wholeness, the ultimate network in which all things are connected.

Just as particles are entangled, so too are we connected to one another, to nature, and to the divine. The connections revealed by entanglement point to a universe guided by balance and interdependence, reflecting the intentions of the God Mind.

The universe is not static; it is alive with movement and interaction. Every thread of the cosmic network contributes to the design, but the God Mind is the unifying presence that ensures coherence.

The God Mind doesn't micromanage but responds to the ripples within the system, allowing creation to unfold while maintaining harmony.

Love, as the highest frequency of alignment, could be seen as the energy that binds the threads of the network, ensuring that connection becomes creation rather than chaos.

Through entanglement, we glimpse our relationship with the God Mind not as separate creations but as threads in the same living system. This perspective calls us to recognize unity and that every action we take affects the whole, reminding us that we are part of something greater.

Just as the weaver guides the loom, we are invited to align with the divine design by choosing actions that resonate with balance and love.

If the universe is a web, then the God Mind is both its source and its sustainer. Every thread, every connection, every ripple through the network is an expression of its energy and intention. To align with this design is to embrace our role in the cosmic tapestry, co-creating a pattern that reflects the beauty and harmony of the divine.

While quantum entanglement operates at the subatomic level, its implications ripple into human experience. At every scale, the universe is relational; an interplay of connections that give rise to patterns, systems, and meaning. For humanity, these connections are most evident in bonds between individuals that creates ripples of influence that shape communities, cultures, and the world.

Humanity's relationship with the environment mirrors the interdependence revealed by entangled particles. Moments of empathy, intuition, and spiritual connection suggest that the human mind participates in a larger field of awareness, where unity transcends individual boundaries.

Entanglement, in this broader sense, becomes a metaphor for the interconnected web of existence we inhabit every day. The bonds we form with others can be thought of as resonances within the human web:

Love and Compassion: These emotions create coherence, much like harmonic frequencies aligning in a system. Acts of kindness, understanding, and care strengthen the threads of connection.

Conflict and Fear: These create dissonance, introducing imbalance into the system. Just as entangled particles reflect the state of their partners, human relationships mirror the energies we project into them.

This relational dynamic echo the principles of entanglement: the state of one part influences the whole, emphasizing the importance of intentional, compassionate action.

Human experience often reveals moments that feel inexplicably connected, as if we are part of a shared energy field. The ability to sense another's emotions reflects a deep resonance between individuals, akin to the shared state of entangled particles.

Insights that seem to arise "out of nowhere" suggest a participation in a broader field of awareness, where the boundaries between self and other blur. Encounters with the divine often carry a sense of unity, as if touching the God Mind directly connects us to the entire web of existence.

While these experiences are subjective, they align with the

relational truths revealed by entanglement: we are not isolated beings but threads in a larger, interwoven fabric.

Understanding ourselves as part of a vast web of connection carries profound implications.

Just as particles influence one another across distances, our actions ripple outward, affecting others in ways we may never see. Acts of love, compassion, and stewardship strengthen the connections within the network, while selfishness and harm weaken it.

Recognizing our role in the cosmic web invites us to live with intention, aligning our actions with the principles of balance, harmony, and care.

This is not a metaphorical responsibility; it is a practical one. The relationships we nurture and the energies we contribute have real consequences for the world we inhabit.

If the universe is a web of connection, then our lives are lived within it. Every thought, every action, every relationship becomes part of the larger design.

To live with awareness of this entanglement is to recognize that:

We are not alone. We are threads in a greater whole, united by relationships that transcend physical boundaries.

Our choices matter. They ripple outward, influencing the pattern of the web in ways both profound and subtle.

Our purpose is participation. By living with love, compassion, and intention, we contribute to the beauty and balance of the cosmic design.

As we've seen, the universe operates as a vast web of connection. Entanglement is not just a phenomenon of particles; it's a harmony that echoes through nature, relationships, and the very fabric of existence. But this web of connection is not static. It is dynamic, alive, and participatory, calling humanity to recognize its role within the cosmic design.

Understanding ourselves as threads in the universal web carries profound implications:

Participation, Not Isolation: Every thought, word, and action sends ripples through the web, influencing outcomes far beyond our immediate awareness.

Alignment with the Whole: To live with awareness is to tune ourselves to the harmony of the larger system, acting with love and intention.

The Ripple Effect: Small, conscious actions like the acts of kindness, stewardship, and creativity can transform the entire pattern, strengthening the web for generations to come.

DNA: A Personal Thread in the Cosmic Tapestry

Within this web, each of us carries a unique thread of connection in the form of DNA, a biological record of our origins and adaptations. DNA encodes not only the instructions for building our bodies but also traces of our shared past, from survival strategies to ancient migrations.

These genetic connections link us to our ancestors and to one another, forming a biological and quantum network that transcends time.

What if DNA is more than a blueprint? Could it act as a repository of memory, storing not only physical traits but echoes of the lived experiences of those who came before us? This idea suggests that the web of connection extends not just across space but through time, entangling us with the stories of our past.

The concept of DNA memory opens profound questions:

Could the shared experiences encoded in our DNA offer glimpses of humanity's collective story, connecting us to distant parts of our past?

Might this entanglement of memory explain phenomena like ancestral intuition or the sense of familiarity with places, cultures, or people we have never encountered?

While these ideas remain speculative, they align with the relational truths revealed by entanglement. The web of connection may not only bind us to the present but to the past and future as well, inviting us to see ourselves as part of a continuum far greater than the self.

Recognizing our place in this interconnected web calls us to live with intention:

Stewardship: Our actions today shape the threads that future generations will inherit, whether through environmental care or the energies we contribute to the human story.

Awareness: By understanding the ripple effects of our choices, we can act in ways that strengthen the web rather than weaken it.

Unity: The realization that we are part of a shared story woven through DNA, relationships, and the cosmos offers a profound sense of belonging and purpose.

The universe calls us not to observe but to participate. To live with awareness of the web is to embrace the profound truth that we are connected across space and time, through relationships, energy, and the memories encoded in our very being.

Every action contributes to the pattern, shaping not only the present but the future.

By aligning with love, compassion, and purpose, we become conscious participants in the unfolding design of creation.

Entanglement reveals that the universe is not a collection of isolated parts but a living tapestry of connection, where every thread is vital to the design. From the quantum realm to the human experience, this web invites us to see our place within the greater whole not as passive observers but as co-creators. Each thought, each action, each ripple we create becomes part of the divine algorithm, shaping a reality alive with possibility, purpose, and love.

To live within this web is to honor the threads that bind us, to recognize the sacred unity of existence, and to contribute to the unfolding masterpiece of the God Mind. The universe calls us to weave with care, intention, and joy, knowing that the patterns we create today will resonate through eternity.

CHAPTER 6: THE NATURE OF TIME AND TIMELESSNESS

Time is one of the most profound and puzzling aspects of existence. It shapes how we perceive reality, giving structure to our memories, our actions, and our expectations for the future. We measure it with clocks and calendars, charting the days, months, and years as if they are fixed markers on an endless line. But is time truly what we think it is?

Both science and spirituality challenge the classical notion of time as a linear flow. At the quantum level, particles seem to move beyond the constraints of past, present, and future, interacting as though time is fluid or irrelevant. In spiritual teachings, time is often described as an illusion; a mental construct that obscures the eternal "now."

And yet, while time may be a construct, it leaves traces in the fabric of existence. Every action, every choice, every moment creates an imprint, a ripple that influences what comes next. These traces, the fingerprints of the past, shape the present and offer a map of where we have been, even as the future remains an unfolding mystery.

In this chapter, we will explore the nature of time and timelessness. What does it mean to live in a universe where time may be an illusion? How do the traces of history interact with the eternal "now"? And how might this understanding reshape our perspectives on life, death, and existence itself?

The answers lie not in linear progression but in stepping into the timeless dance of creation, where the present moment becomes the meeting point of all that was and all that will be.

At the quantum level, time begins to behave in ways that challenge our everyday assumptions. Particles don't move in neat, predictable lines from past to future. Instead, they inhabit states of possibility, appearing to transcend the flow of time as we experience it.

Key phenomena that reshape our understanding of time include:

Superposition: Particles can exist in multiple states simultaneously, defying the idea that events occur in a strict sequence.

Entanglement: As discussed earlier, entangled particles influence each other instantaneously, regardless of distance, seemingly bypassing the limitations of time.

Time Reversal Symmetry: Certain quantum equations suggest that time does not inherently flow in one direction. Processes at this scale are reversible, implying that the distinction between "past" and "future" may be a matter of perspective.

In quantum mechanics, the concept of the observer is not about a person passively watching events unfold. Instead, it refers to any interaction that forces a quantum system to take on a definite state. This could be a human measurement, a particle collision, or even an interaction with a field. The observer "locks in" a moment of potential, transforming it into reality as we perceive it.

The role of the observer shows that the universe is not a static backdrop but an active, participatory system. Observation creates a trace, anchoring possibilities into the flow of experience. The observer is not limited to human awareness. It includes all interactions within the quantum field, suggesting that the universe itself is a kind of observer, shaping and being shaped by its own processes.

This interplay between observation and time suggests that linear time is a tool for organizing the traces left by the dynamic interactions of the quantum field. What we perceive as the flow of time may emerge from this constant process of potential becoming reality.

Quantum mechanics suggests that time at its core is less like an arrow and more like a dance where the future may influence the past just as the past influences the future, creating a feedback loop rather than a one-way stream.

The collapse of the wave function occurs in the "now," the moment where possibilities become reality. This suggests that the present is not merely a point in time but the very fabric of creation.

Time in the quantum realm forces us to question whether linear time is an illusion constructed by the human mind. If all moments exist in a timeless state of potential, then what we experience as time may be the trace left by the unfolding of these possibilities into reality.

As we move forward, we will explore how this quantum perspective aligns with spiritual insights and the ways it challenges our understanding of life, death, and existence itself.

The Illusion of Linear Time

For most of us, time feels like a one-way street, a steady flow from past to present to future. We structure our lives around this perception, marking milestones, measuring progress, and planning for what lies ahead. But is this forward march of time a fundamental truth or a construct of human experience?

Physics offers clues that linear time may be more illusion than reality. Albert Einstein's theory of relativity showed that time is not absolute. It bends and stretches depending on speed and gravity. A clock on a speeding spaceship runs slower than one on Earth, and time near a massive black hole flows differently from time in open space.

Some physicists propose that all of time, past, present, and future, exists simultaneously, like a frozen block. What we experience as the "present" may be a subjective slice of this greater whole but in the quantum world, particles exist in multiple states at once until observed, defying the linear progression we associate with time.

These findings challenge the classical view of time as a rigid sequence of events, suggesting instead that it is a flexible and relational phenomenon.

Our perception of time is deeply tied to how our minds process experience, the past exists as memory, and the future as anticipation. Both are constructs of the mind, giving us the sense of movement through time.

Neurological studies show that even our perception of the present is a composite; a blend of sensory data processed slightly after the fact. What we think of as "now" is a mental reconstruction.

Time, then, may be less a fundamental property of the universe and more a framework created by the mind to make sense of change.

Long before physics questioned the nature of time, spiritual traditions spoke of its illusory nature:

Buddhism: Time is seen as a mental construct. The only true reality is the present moment, where enlightenment can be found.

Hinduism: The cycles of samsara (birth, death, and rebirth) emphasize the eternal nature of existence, transcending linear time.

Christian Mysticism: Teachings often describe God as existing outside of time, with eternity encompassing all moments as one.

These perspectives align with the scientific view that time as we experience it is not the ultimate truth but a relative and subjective phenomenon.

If linear time is an illusion, what does that mean for how we live? It suggests that freedom comes from letting go of attachments to the past and anxieties about the future, focusing instead on the present moment. Recognizing that time is not a line but a continuum allows us to approach life with a broader perspective, embracing the eternal "now."

If time is a construct and linearity is an illusion, then what remains? Spiritual teachings and quantum mechanics alike point to a profound truth: the present moment, the now, is the foundation of reality. The "now" is not just a fleeting point in time but the stage where potential becomes actual, where the dance of creation unfolds.

The collapse of the wave function occurs in the present. A particle's state is undefined until the moment it is observed, suggesting that the "now" is the point where possibilities crystallize into reality.

Teachings across traditions describe the present moment as sacred. In Christianity, Jesus proclaims, *"The kingdom of God is within you"* (Luke 17:21), emphasizing that divine connection is found in the "now." Buddhism and Taoism similarly stress the eternal present as the path to enlightenment and harmony.

The "now" is not merely a slice of time, it is the eternal field where all creation takes place.

The present moment transcends the boundaries of time. The "now" carries the traces of the past, shaping the foundation upon which new possibilities arise. It also holds the seeds of the future, as intentions and actions in the present influence what unfolds.

Time's fragmentation dissolves in the present, allowing us to experience the unity of existence. This is why moments of deep connection, awe, or spiritual insight often feel timeless.

Living in the eternal now means stepping out of the stream of time and into the infinite potential of the present. To live in the "now" is to find freedom from the illusions of time.

The past exists only as memory. By focusing on the present, we free ourselves from the weight of regret and judgment. The future is a field of possibility, shaped by the choices we make now. Grounding ourselves in the present brings clarity and calm. Each moment is a new beginning, a chance to realign with purpose and intention.

The "now" is where we create, heal, and grow. The eternal now is also the space where humanity meets the God Mind. In the "now," we are most attuned to the flow of divine energy, aligning ourselves with love, compassion, and purpose.

The "now" is where we participate in the divine algorithm, shaping reality through our thoughts, intentions, and actions. We create a sacred space where the present moment becomes a temple; a meeting place where the infinite touches the finite, and the timeless interacts with time.

To live in the eternal now is to embrace the heart of existence. It is to see each moment as an opportunity:

To create with intention.

To connect with the divine.

To live fully, free from the illusions of time.

The "now" is not something we pass through, it is the very essence of life. By stepping into the eternal now, we align ourselves with the flow of creation, becoming conscious participants in the timeless dance of the universe.

Jonah's story, found in the Book of Jonah, offers a profound exploration of time, transformation, and the eternal now. Far from being a simple tale of disobedience and correction, it illustrates how the past leaves its trace, the present holds infinite possibilities, and the future is shaped by intentional action.

Time is often seen as unchanging in its flow, but Jonah's story challenges this notion. The past, though fixed in its trace, does not dictate the future. Jonah's initial resistance to God's call to go to Nineveh (Jonah 1:1-3) left its mark, a decision to flee rather than face his fears. Yet, his story didn't end there. The eternal "now" remained open, a space where change could unfold.

Inside the belly of the fish (Jonah 1:17-2:10), Jonah experienced a timeless pause. It was a moment for reflection and surrender, where he prayed:
"When my life was ebbing away, I remembered you, Lord, and my prayer rose to you, to your holy temple" (Jonah 2:7).

This pause illustrates the transformative power of the eternal now, a sacred space where the trajectory of one's life can be redirected through alignment with divine purpose.

When Jonah finally fulfilled his mission, delivering God's warning to Nineveh, the city faced its own pivotal moment (Jonah 3:4-10). The people of Nineveh could have ignored the prophecy or dismissed it as inevitable doom. Instead, they acted in the present, repenting and turning from their destructive ways:
"When God saw what they did and how they turned from their evil ways, he relented and did not bring on them the destruction he had threatened" (Jonah 3:10).

This choice redefined Nineveh's future, demonstrating that prophecy is not about fixed outcomes but about possibilities. It is a dialogue between the divine and the human, a call to align with the eternal now to shape a better trajectory.

Jonah's story reminds us that the future is not a predetermined path but a field of possibilities, shaped by present actions. Just as God's warning to Nineveh was conditional, so too are many moments in life:

"Do this and you will be blessed; do not, and you will face the toll of your actions" (Deuteronomy 28).

This interplay mirrors the quantum principle of wave function collapse, where the future exists as probabilities until a choice crystallizes one into reality. Nineveh's decision to repent shifted the city's trajectory from destruction to renewal, revealing the profound responsibility inherent in the now.

Jonah's narrative invites us to reconsider how we live within the framework of time. His journey—from resistance to realignment, from despair to purpose—reveals that transformation is always possible. The present moment is not just a fleeting point in time but the crucible where past, present, and future converge.

Nineveh's collective repentance reminds us that the eternal now holds the power to:

Heal the Trace of the Past: Choices made in the present can transform the outcomes of what came before.

Redefine the Future: By aligning with divine will, we co-create a future filled with renewal and possibility.

Embrace the Opportunity of the Moment: The now is a sacred space where transformation begins.

The story does not end with Nineveh's transformation; it continues with Jonah's own final lesson (Jonah 4:6-11). Struggling with frustration and a lack of understanding, Jonah retreats, only to be confronted by God's gentle teaching through the incident of the bush:

"You have been concerned about this plant, though you did not tend it or make it grow... Should I not have concern for the great city of Nineveh?" (Jonah 4:10-11).

The bush symbolizes the fleeting comforts of life and calls Jonah's gratitude into check. True transformation requires not only external obedience but internal alignment with love and joy for divine purpose.

Jonah's journey reflects the interplay of time, choice, and divine will. It shows us that the past leaves its trace but does not define the future. The eternal now is a space of infinite possibility, where transformation unfolds and that gratitude and alignment are keys to fully embracing divine purpose.

By stepping into the eternal now, we recognize that life is not bound by the regrets of the past or anxieties about the future.

Instead, it is a sacred opportunity to co-create with the God Mind, shaping a trajectory of renewal and growth.

A Timeless Universe

When we step back and view the cosmos as a whole, time begins to dissolve as an absolute framework. From the quantum to the cosmic, the universe reveals itself as a living, timeless system where past, present, and future are not rigidly divided but intricately woven together.

Time, rather than being a universal constant, appears as a relational construct, a way of organizing change and motion within the fabric of existence. But beyond this framework lies a timeless reality, a universe alive with possibilities and interactions that defy the limits of linear thought.

Einstein's theory of relativity offers a glimpse into this timelessness. In the relativistic view, time is not fixed but depends on an observer's position, motion, and proximity to massive objects. This makes time a localized phenomenon rather than a universal constant and what we perceive as "now" is simply our personal slice of a greater reality.

This challenges our everyday assumptions, inviting us to see time not as a linear arrow but as a dynamic field where all points are connected.

On both cosmic and microscopic scales, the universe is a master of transformation. Stars burn their fuel, collapse, and recycle their elements into new generations of stars, planets, and even life itself. This cycle mirrors the microcosmic processes within our own bodies: cells die, renew, and reform, constantly recreating the structures that sustain us.

Even at the quantum level, particles interact, exchange energy, and reform, maintaining the balance of existence while allowing for infinite possibilities. This interplay of destruction and creation reflects the essence of life and death, not as opposites, but as partners in the continuous dance of existence.

In this timeless universe, the eternal now becomes the canvas where creation unfolds. Each moment is a meeting point where the traces of the past influence what is possible and the potential of the future collapses into reality.

The present becomes a space of co-creation, where divine intention and human action shape the flow of existence.

This interplay mirrors the quantum realm, where superpositions collapse into definite states through observation. The universe itself could be seen as an observer, constantly transforming potential into the ever-unfolding now.

While we observe these cycles with increasing clarity, the mystery of alignment with the God Mind remains. Biblically, there are hints that the spirit world, where energy and essence are preserved, may have a physical aspect beyond our current comprehension. What is evident, however, is the desire of the God Mind to preserve the memories and lessons of creation.

This preservation, described as the great "book of life", represents a continuity of purpose. Those aligned with God's will are not lost to the chaos of separation but are integrated into the grand learning process of the God Mind. It suggests that while the universe recycles matter and energy, the essence of alignment transcends these processes, contributing to something eternal.

At the smallest scales, this principle of preservation and renewal is evident. DNA, for example, carries the memory of life across generations, encoding not only survival traits but the history of existence itself. Even the mitochondria within our cells, once independent organisms, have become integral parts of the human body, preserved and transformed to serve a collective purpose.

Similarly, the brain's neural networks demonstrate a microcosmic example of alignment and preservation. Synaptic connections that align with purpose and learning are strengthened, while those unused or misaligned are pruned away; a micro-level reflection of the divine process of keeping what is meaningful and releasing what no longer serves.

A timeless perspective offers another lens on how we can look at death and transformation:

Death as a Shift: In a timeless universe, death is not an end but a transition, a reorganization of energy and matter within the eternal now. The traces of a life lived remain, influencing the present and future in profound ways.

Legacy as Imprint: Just as the past leaves its trace in the present, our actions and choices ripple outward, becoming part of the cosmic web that future generations inherit.

Life and death are not opposites but phases of a continual process, reflecting the timeless flow of creation and transformation.

What remains beyond our understanding is how alignment with the God Mind manifests in its fullest sense. Is there a physical aspect to this spirit world, or is it purely an energy state that transcends material form? The Bible hints at a realm where energy and essence are preserved for the collective future, but the mechanics of this remain veiled. What we do perceive is the God Mind's intent to preserve its memories and its will as part of a grand learning process.

Our greatest certainty is the promise that what aligns with God's will is preserved. This alignment, marked by gratitude, love, and intentionality, ensures that individuals and actions contribute to the continuity of the God Mind. The "book of life" becomes not just a record but a living testament to the unity of creation, holding within it the essence of those who reflect the divine purpose.

To live in a timeless universe is to recognize the now as the meeting point of all that was and all that will be and to understand that every action creates ripples, influencing the future and reflecting the traces of the past. If we trust the flow we align with the God Mind's perspective, finding peace in the unfolding design rather than clinging to the illusion of control.

A timeless universe is not static, it is alive with possibility, motion, and connection. The eternal now becomes the heartbeat of existence, calling us to step into the flow and co-create with intention, trust, and love. By embracing this timeless reality, we align ourselves with the divine rhythm of the cosmos and the infinite potential of the present moment.

Time, as we experience it, is both a mystery and a marvel. Whether perceived as an arrow moving from past to future, a relational construct shaped by observers, or an eternal now where all possibilities converge, it remains the thread upon which the tapestry of existence is woven.

But this thread is not fixed. The quantum world shows us that time is malleable, relational, and deeply connected to observation and interaction. Spiritual teachings reveal it as an illusion, a tool for understanding the infinite dance of creation. Together, these perspectives invite us to see time not as a limitation but as a dynamic and participatory element of the universe.

In this timeless reality:

The past leaves its trace, but it does not dictate the future.

The present is a canvas of infinite possibility, where transformation begins.

The future is shaped by intentional actions, rippling outward to create the unfolding story of existence.

To live in a timeless universe is to recognize that the "now" is all we truly have. It is the space where we create, connect, and align with the God Mind's flow. By stepping into the eternal now, we become participants in the divine algorithm, co-creators of a reality that reflects love, purpose, and unity.

As we move forward, we'll explore how these threads of connection extend to the deepest aspects of life itself, inviting us to see the universe not only as a place of endless wonder but as a reflection of the divine in motion.

CHAPTER 7: THE IMPRINT OF THE DIVINE

In every corner of the universe, from the most distant galaxies to the smallest cells within our bodies, there is a profound sense of order and purpose. Patterns emerge that are far too intricate and interconnected to be the result of mere randomness. These patterns speak of intention, structure, and coherence; an imprint woven into the very fabric of existence.

Physicists describe the universe as permeated by quantum fields, invisible frameworks that govern everything from the behavior of particles to the creation of stars. These fields are not chaotic; they are precise, finely tuned to allow for the emergence of life and complexity. Biologists uncover similar marvels in DNA, the molecular code that carries the instructions for life. It is a language of four letters, adenine, cytosine, guanine, and thymine, but its expression gives rise to the vast diversity of life on Earth. Across the cosmos and within ourselves, we find echoes of a deeper intelligence, a signature of creation.

This chapter invites us to explore the imprint of the divine within the universe. It is not limited to the observable, it extends into the abstract, into the principles and patterns that govern existence. Quantum fields and DNA are more than mechanisms; they are messages, revealing the fingerprints of the God Mind.

But what does it mean for creation to bear the imprint of the divine? To understand this, we must look beyond the surface and see how these signatures resonate with both scientific discovery and spiritual insight. The patterns of the universe may reflect not only physical laws but also the intention of something far greater, a cosmic algorithm written into existence itself.

We will delve into the quantum fields that sustain reality, the DNA that connects all life, and the spiritual teachings that describe humanity as made in the image of God. Through this journey, we aim to understand not only the divine imprint on creation but also our place within this masterpiece. To carry the imprint of the divine is not just to exist, it is to participate in creation, to co-create with intention, and to align with the flow of the universe itself.

At the heart of existence lies the quantum field, a foundational framework that sustains and shapes reality. Quantum fields are not bound by the physical structures they give rise to; instead, they act as invisible scaffolding, permeating all of space and time. These fields govern the behavior of particles, guide energy exchanges, and provide the backdrop for creation itself.

Among these, the Higgs Field is a striking example. This quantum field gives particles their mass, enabling the formation of matter as we know it. Without this field, the universe would be a sea of formless energy, incapable of producing stars, planets, or life. The Higgs Field and others like it reveal a universe that is not chaotic but precisely tuned, a system with order and coherence that suggests intent.

The quantum world often appears random and unpredictable at first glance, but beneath the surface, patterns emerge:

Wave-Particle Duality: Light and matter behave both as particles and waves, demonstrating the duality embedded in creation itself. This interplay allows for both form and flow, structure and movement.

Quantum Entanglement: Particles separated by vast distances remain mysteriously connected, acting as if they are part of the same unified system.

This suggests that the universe operates not as isolated parts but as an interconnected whole. These patterns reflect a coherence that extends beyond the material. They hint at an underlying order, as if the universe carries a blueprint, a quantum signature, that holds it all together.

What drives this coherence? The principles that govern quantum fields can be thought of as part of a divine algorithm, an intricate set of rules that guide the unfolding of creation. This algorithm:

Balances Chaos and Order: The quantum realm allows for randomness, but this randomness is bounded by principles that maintain stability and allow for the emergence of complexity.

Supports Evolution: Through interactions, quantum fields create opportunities for growth, change, and adaptation, mirroring the dynamic process of life itself.

Reflects Intention: The fine-tuning of the universe's constants, the strength of gravity, the speed of light, the charge of the electron, suggests not randomness but precision.

In this way, quantum fields reveal not only the mechanisms of reality but also its meaning. They are not merely forces, they are signatures of the God Mind, encoded in the fabric of existence.

While quantum fields sustain the universe at its most fundamental level, their effects ripple upward, influencing the formation of matter, the emergence of life, and the intricate processes that sustain it. From the subatomic scale to the cosmic, these fields connect everything in a seamless web of interaction.

As we move forward, we will explore how this divine signature extends into the biological realm, where DNA carries the imprint of creation in the language of life. Together, these layers of reality, quantum and biological, form a unified testament to the divine intention woven into the universe.

Within every cell of your body lies a molecule of profound complexity and beauty: DNA. This double-helix structure is far more than a chemical compound, it is a biological archive, a living code that carries the instructions for building and sustaining life. Encoded within its sequences of adenine (A), cytosine (C), guanine (G), and thymine (T) are not only the traits of individual organisms but the history of evolution itself.

DNA is often referred to as the "blueprint of life," and rightly so. Its structure allows for the replication of life across generations, enabling the continuity of species while also allowing for adaptation and change. But is DNA merely a molecular machine, or could it also carry a deeper imprint, a divine signature that links us to the God Mind?

DNA operates like a language, using a four-letter alphabet to encode the instructions for life. This language is:

Highly Efficient: Each strand of DNA can store vast amounts of information, with the human genome containing over three billion base pairs.

Universally Understood: The genetic code is nearly universal across all life forms, from bacteria to humans, pointing to a shared origin and interconnectedness.

Dynamic: DNA can mutate, adapt, and respond to environmental pressures, allowing for the evolution of life in a way that mirrors the creative flow of the universe.

In this sense, DNA is not only a biological mechanism but also a testament to the creativity and intelligence embedded in creation itself.

It is as if the God Mind inscribed the language of life into the fabric of existence, allowing it to evolve, adapt, and flourish.

Beyond encoding traits, DNA carries a form of Ancestral Memory: DNA connects us to our ancestors, preserving their genetic contributions while also carrying the marks of their experiences, such as adaptations to environmental challenges.

Recent discoveries in Epigenetics show that environmental factors can influence gene expression, turning certain genes on or off in ways that can be inherited by future generations. This suggests that DNA is not static but responsive, a living record that reflects the ongoing story of life.

This capacity for memory aligns DNA with the concept of the eternal now, where the traces of the past influence the present while shaping the future. It is a biological expression of the timeless flow of creation.

If quantum fields represent the foundational framework of the universe, DNA serves as the bridge between the physical and the divine. Through its intricate design, DNA connects life to the Cosmos.

The elements that form DNA, carbon, hydrogen, oxygen, nitrogen, and phosphorus, were forged in the hearts of stars, linking the molecular basis of life to the cosmic processes that sustain the universe.

The structure, function, and adaptability of DNA reflect a higher intelligence, a creative force that imbues life with purpose and direction. The universality of the genetic code highlights the interconnectedness of life, mirroring the relational nature of the God Mind.

DNA, like quantum fields, reveals the signature of the divine. Its patterns, processes, and potentialities point to a reality where life is not an accident but a profound expression of creation.

Its language reflects the precision of the divine algorithm. Its adaptability demonstrates the dynamic nature of creation. Its interconnectedness reveals the unity of all life, a reflection of the God Mind's intention for harmony and coherence.

As we continue, let's explore how these scientific insights align with spiritual teachings, where the Word, the Logos, is seen as both the source and sustainer of life.

In the opening verse of the Gospel of John, we find one of the most profound declarations in spiritual literature:
"In the beginning was the Word, and the Word was with God, and the Word was God" (John 1:1).

The "Word," or Logos in Greek, is more than spoken language.

It represents the divine principle of order, reason, and creative power. The Logos is the organizing force that brings coherence to the cosmos, transforming chaos into creation. It is the blueprint, the vibration, and the intention that manifests the universe.

The Gospel continues:
"The Word became flesh and made his dwelling among us" (John 1:14).

This declaration bridges the divine and the material, suggesting that the Logos is not distant or abstract but intimately connected to creation. The Word made flesh reveals that the divine essence can be embodied, manifesting in the physical realm through life, action, and purpose.

Just as the Logos is a language of creation, DNA serves as the language of life. Both are systems of communication, encoding and transmitting the patterns that sustain existence. The quantum fields that underlie reality can be seen as the medium through which the Logos operates, shaping energy into form and potential into reality.

The idea of the Word as a creative principle resonates with teachings across cultures:

Hinduism: The concept of *Om*, the primordial sound, is described as the vibration from which the universe originates. Like the Logos, it represents both the source and the essence of creation.

Indigenous Wisdom: Many Indigenous traditions speak of the world being sung into existence, where sound and intention are inseparable from the act of creation.

Modern Science: Theoretical physics increasingly explores the role of vibrations and frequencies in shaping the universe, aligning with ancient insights about the creative power of sound and intention.

These parallels suggest that the Word is not bound by a single tradition but reflects a universal truth, the presence of a divine principle at the heart of existence.

In a timeless and interconnected universe, the Word functions as a divine algorithm, a set of principles that guide the unfolding of creation. Like DNA and quantum fields, the Logos is not rigid but responsive, allowing creation to evolve in harmony with divine intent.

The Logos provides structure while leaving room for free will and creativity, reflecting the balance between divine guidance and human agency.

The Logos unites all of existence while allowing for infinite expressions, mirroring the interconnectedness of life. Through the Word, the universe becomes a living system, where every part reflects the whole, and every action contributes to the divine pattern.

To align with the Word is to embody the principles of creation in our own lives:

Speaking with Intention: Recognizing the power of words to shape reality, both for ourselves and others.

Creating with Purpose: Using our actions, choices, and creativity to contribute to the greater harmony of existence.

Embodying the Divine: Living as reflections of the Logos, carrying the imprint of the divine in our thoughts, words, and deeds.

The Word made flesh is not just a historical event, it is an ongoing process, where the divine continues to manifest through humanity and the cosmos.

The Logos connects the quantum and the spiritual, the material and the divine. It is the thread that weaves through the quantum fields, DNA, and the human experience, revealing a universe alive with purpose and coherence. To see the Word in creation is to glimpse the God Mind itself, a timeless intelligence calling us to participate in the unfolding story of existence.

The Divine Algorithm

An algorithm is a set of instructions or rules that guide a process toward an intended outcome. In the context of the universe, we see a divine algorithm, a framework of principles, patterns, and interactions that sustain and evolve existence. This divine algorithm operates at every scale, from the quantum to the cosmic, reflecting the coherence and intelligence of the God Mind.

At the smallest scales, quantum mechanics demonstrates precision and adaptability. Particles interact according to probabilities, yet within these interactions, patterns of order emerge, allowing for complexity and stability.

DNA, as explored earlier, is a biological algorithm, encoding the instructions for life while remaining dynamic and responsive to change.

The fine-tuning of the universe's constants, gravity, electromagnetism, the speed of light, suggests a precise balance that allows for the emergence of stars, planets, and life itself.

Together, these layers of reality form a living algorithm, one that guides the unfolding of creation with intention and purpose.

The divine algorithm is not rigid or deterministic. It balances chaos and order, allowing for creativity and freedom within the boundaries of coherence. Randomness and uncertainty provide the raw material for innovation, enabling the universe to explore infinite possibilities.

Patterns and principles ensure that this creativity does not devolve into chaos, sustaining the harmony of existence.

This interplay mirrors the relationship between free will and divine guidance. The divine algorithm provides the framework, but within it, there is room for choice, growth, and evolution.

The divine algorithm operates through the language of creation, evident in the mathematical precision of the universe, from Fibonacci sequences in nature to the constants that govern physics, reflecting an underlying logic.

Recurring patterns, such as fractals, reveal a self-similar design across scales, echoing the idea that creation is unified yet infinitely diverse. At its core, the universe functions as a dynamic exchange of energy and information, a flow that connects all aspects of existence. This language is not static, it is alive, evolving with the unfolding story of creation.

To align with the divine algorithm is to live in harmony with the principles of creation, recognizing the patterns and rhythms of the universe and aligning our actions with them.

Participating in the process of creation, contributing to the beauty and coherence of the greater whole. Celebrating the infinite expressions of life while honoring the unity that connects all things.

Alignment is not about perfection but about participation, being an active part of the divine process, living with awareness, purpose, and gratitude.

The divine algorithm also offers a profound perspective on human potential. Just as the universe evolves, so too do individuals. Each moment is an opportunity to learn, adapt, and grow in alignment with divine purpose.

Humans, as bearers of the divine imprint, are invited to co-create with the God Mind, shaping reality through their choices and actions. The traces we leave, our words, actions, and creations, become part of the divine algorithm, influencing the future in ways we may never fully see.

This perspective transforms life into a sacred journey, where every decision contributes to the unfolding masterpiece of creation.

The divine algorithm is not a static formula, it is a living process, a dynamic interplay of energy, intention, and action. To see the universe through this lens is to glimpse the intelligence of the God Mind at work, guiding creation with grace and purpose. By aligning ourselves with this algorithm, we become active participants in the story of existence, co-creators in the timeless dance of the cosmos.

The Divine Imprint on Humanity

The concept that humanity is made "in the image of God" (Genesis 1:27) speaks not of physical likeness but of something far greater: a shared essence, a divine imprint carried within every human being. To live in the image of God is to reflect the creativity, love, and purpose that flows from the God Mind, participating in the unfolding story of creation.

Humanity's divine imprint is evident in:

Creativity: Just as the God Mind creates the cosmos, humans are gifted with the ability to create art, ideas, relationships, and solutions. Our creativity mirrors the divine process of bringing potential into reality.

Consciousness: The human capacity for self-reflection, empathy, and moral reasoning sets us apart as participants in the divine algorithm, capable of aligning our will with the flow of creation.

Interconnection: Like the quantum fields and DNA that connect all life, humans are deeply interconnected, both with one another and with the universe. This relational nature reflects the unity of the God Mind.

To carry the divine imprint is both a gift and a responsibility. It calls us to:

Align with Purpose: Recognize and embrace our role as co-creators, acting in ways that nurture life, beauty, and harmony.

Live with Gratitude: Appreciate the profound gift of existence, seeing every moment as an opportunity to contribute to the divine design.

Uphold Unity: Strive to bridge divisions, honor diversity, and recognize the sacredness of all life.

Living in the image of God means seeing ourselves not as separate or superior but as integral parts of the greater whole.

While the divine algorithm provides a framework for harmony, free will allows for choices that disrupt this balance. Acts of greed, hatred, and division pull us out of alignment with the God Mind.

Despite these divergences, the divine imprint within us remains, continually calling us back to love, compassion, and purpose. This tension is not a flaw but a feature of the divine design, inviting humanity to grow, learn, and choose alignment consciously.

Living in the image of God is an invitation to become co-creators, shaping reality with intention and care:

Creating Beauty: Whether through art, relationships, or acts of kindness, humans have the power to bring light into the world, reflecting the creative essence of the God Mind.

Healing the Web: By recognizing our interconnectedness, we can work to heal the divisions and imbalances that disrupt the harmony of creation.

Living with Intention: Every choice, no matter how small, contributes to the unfolding story of existence. To live intentionally is to honor the divine imprint within us.

The imprint of the divine is not confined to the grand scale of the cosmos or the minute intricacies of DNA, it is present in every breath, every choice, and every moment of existence. From the quantum fields that sustain reality to the biological blueprints that carry the history of life, creation speaks a unified language of coherence, beauty, and purpose. This is the divine algorithm in motion, the signature of the God Mind etched into the fabric of all that is.

To recognize this imprint is to awaken to a profound truth: we are not passive observers of creation but active participants. We are bearers of the divine signature, entrusted with the responsibility to nurture, heal, and co-create. The universe, in its timeless wisdom, invites us to align with its flow, to live as reflections of the Logos, and to carry forward the legacy of love, creativity, and unity.

Yet this journey is not without challenges. Free will allows for divergence as much as alignment, and the human story is marked by both harmony and discord. But even in moments of imbalance, the divine imprint remains, calling us back to the grace of the now, where transformation is always possible. Through gratitude, intention, and action, we can honor this imprint, weaving our lives into the greater masterpiece of creation.

As we move forward, let us carry this awareness with us: that every act of love, every spark of creativity, and every effort toward unity reflects the divine image within. To live in the image of God is to embrace the sacredness of existence, to see the divine in all things, and to participate joyfully in the unfolding story of life.

The imprint of the divine is not a relic of the past, it is a living testament to the eternal connection between the God Mind and creation. By embracing it, we step into our role as co-creators, shaping a reality that reflects the infinite beauty and purpose of the universe itself.

The divine imprint is a mirror, reflecting both the God Mind and our potential to align with its flow. To live in the image of God is to embrace this reflection, to see ourselves as carriers of the divine signature, and to act with love, creativity, and purpose. By doing so, we honor the gift of existence and contribute to the unfolding masterpiece of creation.

CHAPTER 8: HUMANITY'S PLACE IN THE WEB OF LIFE

Humanity's place in the web of life is both wondrous and precarious. We exist as the most complex nodes in an intricate network, bridging the material and the metaphysical, the microscopic and the cosmic. Our role is unique, not simply as stewards or dominators, but as participants in a system so vast and interconnected that our smallest actions ripple outward, reshaping the world in ways both seen and unseen.

For millennia, human societies operated in rhythm with nature, guided by instincts, seasonal cycles, and the wisdom of observing ecosystems. Ancient cultures often revered the earth, seeing themselves not as separate from it but as expressions of its vitality. They perceived life as a shared journey, each organism playing a role in maintaining balance. Yet, over time, something shifted. Humanity began to see itself as apart from, rather than a part of, this web, and in that shift, we lost sight of our niche, the unique role for which we were formed.

Every ecosystem is a tapestry of interdependent threads. Plants exhale oxygen; animals inhale it. Predators maintain the balance of prey populations. Even decomposers, such as fungi and bacteria, return nutrients to the soil, completing the circle of life. This intricate dance depends on each participant fulfilling its role within the larger whole.

But humanity has disrupted this balance. From deforestation to industrial pollution, our actions have unraveled many of the natural threads that sustain life. Our technologies, while miraculous, often function in opposition to the harmony found in nature. For example, cyanobacteria once transformed Earth's atmosphere, creating the oxygen-rich world we depend on.

Today, industrial activities release carbon dioxide at rates that overwhelm the planet's ability to maintain equilibrium.

This imbalance, however, is not irreversible. The same ingenuity that created these disruptions can also repair them. Humanity's niche is not simply to take but to create, not to dominate but to harmonize. This potential is encoded in our very essence.

Unlike any other species, humanity carries within it a capacity for self-reflection, innovation, and intentionality. We are not passive participants in the web of life but active agents capable of reshaping it. With this power comes profound responsibility. To find our place within the web, we must first recognize that we are not the masters of creation but co-creators alongside it.

The Hebrew scriptures declare that humanity was formed from the earth's dust, infused with the breath of life. This imagery is echoed in modern science, which shows that every atom in our bodies was once part of something else, a star, a plant, or another living being. We are made of the same substance as the world around us, yet we are something more. Our capacity to reflect, imagine, and innovate is a testament to this distinction.

This duality, being both, of the earth and transcending it, positions humanity as a bridge. We are uniquely equipped to weave meaning into the material world, to transform raw resources into tools, art, and systems that enhance life. But to do so in alignment with the web of life, we must learn to listen to the rhythms of nature, the wisdom of ecosystems, and the silent language of balance that governs the universe.

Reclaiming humanity's niche begins with humility. We must approach the web of life not as engineers seeking to control it but as apprentices eager to learn from its patterns. Nature's design principles, adaptability, interdependence, and resilience, offer a blueprint for restoring harmony.

For example, permaculture practices mimic natural ecosystems to create sustainable agriculture. Reforestation efforts restore habitats and sequester carbon, while renewable energy technologies harness the sun and wind, echoing the age-old reliance on natural cycles. These solutions are not just practical; they are spiritual acts of reconnection. They remind us that our role is not to extract from the earth but to contribute to its flourishing.

Realigning with the web of life requires a shift in perspective. It means seeing ourselves not as separate from creation but as integral to its ongoing story. It means asking not, "What can I take?" but, "What can I give?" This shift is not merely ethical; it is existential. Our survival depends on it.

The indigenous wisdom of many cultures teaches that every action must consider its impact on the next seven generations. This long view aligns with the quantum perspective that every choice resonates through time and space. Just as waves ripple outward, our decisions today shape the reality of tomorrow.

To realign with the web of life is to step into our role as caretakers, innovators, and storytellers. It is to honor the earth not just as our home but as our partner in the grand symphony of existence.

Imagine a world where humanity's presence enhances the web of life rather than diminishes it. Forests flourish, oceans teem with life, and the air hums with the energy of ecosystems in balance. In such a world, human innovation aligns with natural principles, creating technologies that heal rather than harm. This is not a utopian dream; it is a possible future, one that calls us to action.

To find our place in the web of life is to embrace our role as co-creators with the universe. It is to weave ourselves back into the tapestry of existence, not as dominant threads but as vibrant, integral ones. This is our challenge and our calling, to live not as strangers in creation but as its stewards, crafting a legacy of balance, beauty, and belonging.

Humanity's divergence from the natural web of life is not the story of a singular moment, but of countless choices, each one tilting the balance further. These choices are like steppingstones across a river, leading us away from the banks of harmony and into unknown terrain. Some were deliberate, born of curiosity, necessity, or ambition, while others were unintentional, the byproduct of innovation outpacing wisdom. Together, they form the path we now walk, a path both remarkable and perilous.

What set us on this divergent path? Was it the mastery of fire, which allowed us to shape the environment to our will? The cultivation of crops, which tethered us to the land and gave rise to cities? Or was it the Industrial Revolution, with its promise of boundless progress at any cost? Perhaps it was all of these and more, each one a step further from the rhythms of the earth and deeper into a world of our own making.

The story of Cain and Abel is one of humanity's earliest reflections on the divergent responses to our role in creation. These brothers, born into the world outside Eden, represent two paths: one rooted in alignment with divine intention, the other driven by pride and separation. Their choices echo through history, shaping the systems and patterns that continue to influence our world.

Abel's role as a shepherd reflected a life lived in harmony with the systems of nature. He cared for the vulnerable, nurtured balance, and offered the best of his flock to God as a symbol of gratitude and humility. Abel's offering was not merely an act of obedience; it was an acknowledgment of humanity's place within the web of life. His actions aligned with the original command to "tend and keep" the earth, honoring the interconnectedness of creation.

Abel's path exemplifies a response of stewardship and humility, recognizing that humanity's gifts are not for self-glorification but for the flourishing of all. His life is a testament to what it means to live in balance, offering back to the Creator the fruits of a life well-tended.

Cain, in contrast, sought control over nature. As a farmer, he worked the soil, but his offerings lacked the spirit of gratitude and alignment seen in his brother's. Where Abel's sacrifices reflected trust and humility, Cain's offering revealed pride and a desire for self-sufficiency; a declaration that he could thrive apart from divine guidance.

This divergence became more pronounced when Cain's offering was not received as he expected. Instead of reflecting inward or seeking understanding, Cain turned to jealousy and anger, ultimately committing the first act of fratricide. His decision to take his brother's life marked a deeper separation from the web of life, severing the bonds of kinship and community.

Cain's actions reveal the dangers of misalignment. His path illustrates how pride and a desire for dominance lead to cycles of exploitation and destruction. This pattern, of taking without giving and controlling without care, has repeated throughout history, creating systems that rise and fall under the weight of their own imbalance.

Cain's descendants carried forward his legacy, building the first cities and laying the foundations for systems of extraction and control. These systems, while innovative, often prioritized short-term gain over long-term sustainability, mirroring Cain's choices. Across civilizations, we see the echoes of this divergence: empires rising through conquest and falling through excess, societies exploiting the land until it can no longer sustain them.

Yet, even in these patterns of imbalance, the potential for realignment remains. The story of Cain and Abel is not just a cautionary tale; it is a reminder of the choices before us. We can follow Abel's path of harmony, embracing our role as stewards, or we can continue down Cain's path of dominance, risking further disconnection from the divine and from one another.

The story of Cain and Abel offers a mirror for humanity's current crossroads. In a world where the consequences of imbalance are increasingly evident, climate change, resource depletion, social inequities, we are faced with the same fundamental choice: will we live in harmony with creation, or will we continue to dominate and extract until the system collapses?

Choosing Abel's path requires humility, courage, and a willingness to realign. It calls us to see the earth not as a resource to be exploited but as a partner in creation, deserving of care and respect. It invites us to offer the best of what we have not in pride, but in gratitude and to restore the bonds of community and interdependence that Cain severed.

As we reflect on Cain and Abel's divergent paths, we are reminded that the web of life is forgiving but not infinite. The choices we make today will echo through generations, shaping the world we leave behind. Will we walk the path of Abel, tending the garden and honoring the Creator's design? Or will we follow Cain, pursuing control at the expense of balance and harmony?

The answer lies not in the grand gestures of nations but in the daily decisions of individuals. Each choice is a thread in the tapestry, a step along the path. Let us choose wisely for the future of the web of life and our place within it depends on it.

To diverge is not inherently wrong. It is the nature of life to evolve, to adapt, and to explore new possibilities. In many ways, humanity's divergence has been a gift. It has given us art and science, medicine and philosophy, languages that span the globe, and technologies that connect us across vast distances. These achievements are the fruit of our unique capacity for creativity and reflection, a testament to the potential of our role as co-creators in the web of life.

But divergence also comes with burdens. In our quest for mastery, we have often lost sight of balance. Where once we were partners in the dance of life, we became choreographers imposing our own rhythms. This has led to imbalances that ripple outward, from the extinction of species to the warming of the planet. The very systems that sustain life now bear the weight of our divergence, straining under the pressure of unchecked growth and consumption.

Even so, the web of life has not abandoned us. It continues to hold us, offering threads of connection and opportunities to realign.

Consider the resilience of ecosystems, how forests regrow after fires, how coral reefs can recover when given time and care, or how polluted rivers can become clear again.

These natural miracles remind us that balance is not beyond reach. The web of life is forgiving, but only if we are willing to see it again and step back into harmony.

Realigning does not mean erasing our path of divergence. It means finding a way to weave our innovations and insights back into the fabric of life, enriching it rather than unraveling it. It is about asking how our gifts, our creativity, our technologies, our philosophies, can serve the greater whole rather than just ourselves.

We stand at a crossroads, a point where the path of divergence meets the opportunity for reconnection. What lies ahead depends on the choices we make now. Will we continue to walk away from the web, unraveling the delicate threads of interdependence? Or will we turn, not back to the past, but toward a future where humanity and nature thrive together?

This choice is not abstract. It is made every day, in the fields of science and politics, in the classrooms where children learn, and in the homes where families decide what to value and how to live. It is made in the soil, the air, and the water, in every action that either nurtures or depletes the earth. The crossroads is not somewhere out there; it is here, now, in the moment of every decision.

To walk a new path is to become bridges ourselves, connecting the divergent with the harmonious, the material with the spiritual, the old ways with the new. This is not an easy task. It requires humility to admit where we have gone wrong, courage to change course, and wisdom to see that the web of life is not something outside us but something we are part of.

As we step forward, let us carry with us the lessons of both our divergence and our belonging. Let us remember that we are not just observers of creation but participants in its unfolding. And let us walk with the hope that, in weaving our own threads back into the web of life, we can help restore the balance that sustains us all.

History is a living testament to the choices humanity has made, an ever-unfolding narrative of divergence and realignment. In its patterns, we see the echoes of Cain and Abel: cycles of harmony and disruption, flourishing and collapse. Each rise and fall bear the marks of decisions rooted in either balance or misalignment, illustrating the profound interconnectedness of humanity's actions and the web of life.

Across civilizations, the rise of great societies often began with a balance between human needs and the earth's abundance. Ancient agricultural communities flourished by working in harmony with seasonal cycles, developing practices that nurtured the soil and sustained the ecosystem. The terraces of the Andes, the rice paddies of Asia, and the flood-based agriculture of the Nile Valley are examples of humanity's ingenuity operating within natural rhythms.

But as societies grew, so too did their demands on the land. The desire for wealth, power, and control often led to overexploitation, deforestation, and the depletion of resources. These imbalances created cascading effects, soil erosion, water shortages, and the collapse of ecosystems, that contributed to the fall of once-great civilizations.

Consider the fall of ancient Mesopotamia, the cradle of civilization. The once-fertile lands between the Tigris and Euphrates rivers were transformed by irrigation systems that initially brought abundance but later led to salinization of the soil. Crops failed, and the very system that had sustained life became a source of scarcity.

Similarly, the Mayan civilization, known for its advanced knowledge of astronomy and architecture, faced collapse when deforestation and agricultural overreach led to droughts and food shortages. These echoes of imbalance remind us that even the most advanced societies are not immune to the consequences of misalignment.

Yet history also offers examples of resilience and realignment. After the Dust Bowl of the 1930s, a period of severe soil erosion caused by unsustainable farming practices, communities in the United States adopted conservation techniques to restore the land. The implementation of crop rotation, contour plowing, and reforestation helped heal the scars of overuse, showing that even in the face of ecological devastation, recovery is possible when we choose to realign.

Indigenous cultures around the world have long modeled this wisdom, emphasizing the sacredness of the land and the importance of living in balance. The Haudenosaunee Confederacy's principle of considering the impact of decisions on the next seven generations offers a timeless blueprint for sustainable living. These echoes of Eden remind us that the knowledge of balance has never been lost, it has simply been overshadowed by the noise of divergence.

In the modern era, the stakes are higher than ever. Advances in technology and industry have magnified humanity's impact on the web of life, accelerating both creation and destruction. We now face challenges on a global scale: climate change, deforestation, species extinction, and the depletion of natural resources.

These crises are not isolated; they are deeply interconnected, reflecting the same patterns of imbalance that have echoed throughout history. Yet, they also present an unprecedented opportunity. For the first time, humanity has the tools, knowledge, and global awareness to address these challenges collectively.

We stand at a pivotal moment. Will we allow the echoes of imbalance to define us, or will we listen to the whispers of wisdom that call us back to harmony?

To move forward, we must look back, not with nostalgia, but with reverence. The echoes of history are not merely warnings; they are lessons, reminders of what is possible when humanity lives in alignment with the web of life. They call us to remember the balance of Eden, not as a relic of the past, but as a guiding principle for the future.

As we listen to these echoes, we are reminded that the web of life is not silent. It speaks through the cycles of nature, the resilience of ecosystems, and the wisdom of those who have walked before us. It is up to us to heed its call, to weave the lessons of history into a new narrative of balance and belonging.

The echoes of history do not end with warnings; they invite action. Humanity's trajectory is not fixed, and the web of life, though strained, remains resilient. Realignment is not only possible; it is essential. Yet with the awareness of this need comes a profound responsibility. To realign is not merely to repair what has been broken but to embrace a deeper relationship with creation, one marked by humility, stewardship, and reciprocity.

The story of Nineveh offers one of the Bible's most striking examples of realignment. A city steeped in pride and misalignment, Nineveh seemed destined for destruction. Yet, when confronted with Jonah's warning, the people chose a different path. They humbled themselves, acknowledged their missteps, and turned toward a new way of living.

Nineveh's transformation was not merely external; it was a profound shift of the heart. The people not only ceased their harmful actions but embraced gratitude, repentance, and responsibility. This collective choice spared the city and demonstrated that even the most entrenched patterns of imbalance can be undone when humility and courage take root.

Nineveh's story reminds us that realignment is always possible, no matter how far we have strayed. It calls us to reflect on the power of collective action, how communities, nations, and even civilizations can change course when they choose to align with the rhythms of life and the divine will.

Realignment begins with the courage to change. It requires us to acknowledge where we have gone astray, both individually and collectively. This is not an easy task. The systems of imbalance, built over centuries, are deeply ingrained in our economies, cultures, and ways of thinking. To step away from them is to challenge the very foundations of what we have been taught to value: convenience, consumption, and control.

But courage is not about acting without fear; it is about choosing to act despite it. Just as Nineveh's people turned from their destructive ways, so too can humanity transform its relationship with the earth. The first step is recognizing that realignment is not a return to the past but a forward movement, a journey toward harmony that embraces both ancient wisdom and modern innovation.

The story of humanity's role in creation has often been misunderstood. The biblical command in Genesis to "have dominion" over the earth has sometimes been interpreted as a license to exploit, rather than a call to steward. Yet, when seen through the lens of alignment, dominion becomes a sacred trust, a responsibility to tend and keep the garden of creation, as a gardener tends a vine.

A gardener's work is twofold: pruning what no longer serves and nurturing what is fruitful. Pruning is not destruction; it is an act of care, removing what disrupts balance or hinders growth. Genesis speaks of humanity's role in cultivating and keeping the garden, a task that requires discerning what to nurture and what to let go. Jesus deepens this metaphor, teaching that we must also prune the systemic separations from the God Mind, those habits, beliefs, and systems that sow division, exploitation, and imbalance. Only then can we fully nurture connection and life.

Pruning is an act of humility. It acknowledges that not everything we create or hold onto is aligned with the rhythms of life. It invites us to reflect on what disrupts harmony, be it a system of exploitation, a habit of overconsumption, or a belief in our separation from the divine. In letting go of these disruptions, we make space for new growth, for practices and relationships that sustain life and foster unity.

Nurturing, in contrast, is an act of trust. It recognizes the potential for flourishing within the web of life and aligns our actions with that potential. Aligning with the God Mind means cultivating what brings harmony: regenerative systems, reciprocal relationships, and acts of kindness and gratitude. It means seeing ourselves not as owners of the earth but as caretakers, tending the soil of creation with love and reverence.

In both Genesis and Jesus' teachings, the call to steward is not about ownership but about relationship. It is about aligning with the divine intention for creation and participating in its flourishing. When we prune what no longer serves and nurture what is fruitful, we realign with the rhythms of the God Mind, restoring balance to the web of life.

This shift in perspective transforms how we interact with the web of life, moving from a mindset of extraction to one of reciprocity.

Reciprocity is the practice of giving back in gratitude for what we receive. It is a principle found in ecosystems, where every organism contributes to the whole. Trees provide oxygen, fungi return nutrients to the soil, and rivers carry life to the land. Humanity, too, is called to give back, not only to sustain the systems that sustain us but to honor the interconnectedness of life.

This principle can be seen in indigenous practices of thanksgiving and conservation, where every harvest is accompanied by an offering of gratitude and care. It is reflected in permaculture principles, which teach us to work with nature rather than against it, designing systems that replenish rather than deplete. Reciprocity is not just a practice; it is a mindset, a way of being in harmony with the web of life.

With awareness comes responsibility. Humanity's capacity for reflection and innovation sets us apart, but it also places upon us a unique burden. We are the only species capable of understanding the full impact of our actions on a global scale. This knowledge calls us to act, not out of guilt or fear, but out of love and commitment to the greater whole.

The people of Nineveh understood this. Their transformation did not come from external compulsion but from an internal awakening. They recognized the weight of their actions and chose to change, not just for themselves, but for future generations. Their story challenges us to do the same, to embrace the responsibility that comes with awareness and to align our actions with the flourishing of all.

Realignment is not merely a duty; it is a joy. To live in harmony with the web of life is to experience a deeper connection with the earth, with others, and with the divine. It is to see the beauty in every leaf, every river, every moment of shared breath. It is to know that our lives are not separate threads but part of a tapestry, woven together in a dance of interdependence.

As we take steps toward realignment, we begin to heal, not only the earth but ourselves. The fractures within us, born of separation and imbalance, are mended as we rediscover our place in creation. This is the gift of responsibility: it offers us the chance to become whole again, to live in harmony with the divine rhythm of life.

Realignment and responsibility are not endpoints; they are the beginning of a new chapter in humanity's story. The choices we make today will shape the echoes of tomorrow, creating a legacy of balance, beauty, and belonging. Let us walk this path together, with courage, humility, and joy, knowing that every step brings us closer to the harmony for which we were created.

Healing the web of life begins with the recognition that we are not separate from it. Every thread we restore strengthens the whole, and every choice we make resonates across the tapestry of creation. To heal the web is not merely to repair the damage done but to participate in the ongoing work of co-creation, aligning with the rhythms of life and the divine intention that sustains them.

The web of life thrives on interdependence. No creature, plant, or element exists in isolation; each plays a role in sustaining the whole. Rivers nourish the land, trees anchor ecosystems, and pollinators carry life from flower to flower. Humanity, too, has a role, but unlike other beings, we are uniquely capable of choosing how we fulfill it.

Healing the web begins with embracing this interdependence. It requires us to see ourselves not as separate strands but as woven into the very fabric of creation. This perspective shifts how we approach the challenges before us, from climate change to social injustice, revealing them as interconnected problems requiring interconnected solutions.

Practical steps toward healing the web often mirror the natural processes of renewal. Just as ecosystems recover through cycles of growth and regeneration, so too can humanity restore balance through intentional action.

Allowing natural systems to regenerate by protecting habitats, planting native species, and restoring degraded lands. Efforts like reforestation and wetland restoration are acts of healing that benefit not only the environment but also the communities that depend on it.

Building resilient systems and embracing sustainable practices in agriculture, energy, and urban planning. Permaculture principles, designing systems that work in harmony with nature, offer a blueprint for living in alignment with the earth.

Healing the web also involves restoring relationships, both human and ecological. Indigenous teachings remind us that every harvest must be accompanied by gratitude and care, ensuring that what we take is balanced by what we give back.

Each of these actions is a thread of healing, woven into the larger tapestry of restoration. They remind us that even small choices, what we eat, how we travel, how we engage with others, can contribute to the flourishing of the whole.

The web of life is not only external; it is also internal. To heal the fractures in creation, we must also heal the fractures within ourselves. Separation, from nature, from each other, from the divine, creates wounds that ripple outward, shaping how we interact with the world. Healing these wounds requires reflection, humility, and a willingness to reconnect.

Spending time in nature is not merely restorative; it is transformative. It reminds us of our place in the web and rekindles a sense of wonder and belonging.

Healing the web involves building bridges where division has taken root. Acts of kindness, forgiveness, and community-building are as vital as ecological restoration in mending the fabric of life.

Aligning with the God Mind means recognizing that healing is not a solitary endeavor but a partnership with the Creator. It is an act of faith, trusting that our efforts, no matter how small, are part of a greater design.

Healing the web is not the work of a single generation but a legacy to be carried forward. It is a commitment to future life, an act of hope that the choices we make today will create a world where balance and harmony can flourish.

This legacy begins with intention. Each thread we weave back into the web, each act of restoration, each step toward alignment, becomes a foundation for those who come after us. Just as the echoes of Eden call us to remember the balance we were created for, so too do the whispers of the future invite us to imagine a renewed world.

Healing is not the end of the journey but the beginning of a new chapter. As we mend the web, we are not only restoring what was lost but co-creating what is possible, weaving a tapestry of life that reflects the divine beauty and interconnectedness of all things.

The web of life, intricate and interconnected, is a reflection of deeper truths that extend beyond what we see. Beneath the cycles of nature and the rhythms of existence lies a quantum reality, a field of infinite possibilities where every thread is woven into the fabric of creation. This quantum foundation is not separate from the web of life but its very essence, the unseen framework that holds it together.

As we have explored the themes of alignment and divergence, realignment and healing, we find ourselves drawn closer to this deeper understanding. Just as the web of life thrives on interdependence, so too does the quantum realm operate through connection, entanglement, and resonance. Every particle, wave, and interaction participates in a symphony of existence that transcends time and space, echoing the divine intention behind creation.

The journey through this chapter has been a journey through threads, threads of history, responsibility, and restoration. These threads mirror the very structure of the quantum universe, where every point is connected to every other point, and where the smallest actions ripple outward to shape the whole.

Physics teaches us that the observer influences the observed, collapsing waves of potential into particles of reality. In the same way, our choices influence the web of life, transforming possibilities into lived experience. The act of observing, of choosing, is not passive but participatory. It is a sacred act of co-creation, aligning the physical with the metaphysical, the visible with the unseen.

If the web of life is the garden we are called to tend, then the quantum field is its soil, the fertile ground where the seeds of creation take root. Just as a gardener must understand the needs of the soil to cultivate growth, so too must we understand the quantum principles that underlie existence if we are to align with the God Mind.

Genesis tells us that humanity was formed from the dust of the earth, a poetic acknowledgment that we are both of the material world and beyond it. Quantum mechanics echoes this truth, revealing that matter is not static but dynamic, a dance of energy and information shaped by intention. To heal the web of life is to participate in this dance, to recognize that our every thought, action, and interaction shapes the world around us.

As we step into the next chapter, we will delve deeper into this quantum reality, exploring the mind of God as the ultimate observer, the source of all possibilities. The threads we have followed through this chapter, of stewardship, reciprocity, and restoration, are the same threads that weave through the quantum fabric of existence.

This is not the end of our journey but the beginning of a new understanding. The web of life, the quantum field, and the God Mind are not separate concepts but facets of a single, unified reality. As we continue, we will uncover how these truths converge, illuminating humanity's place in the cosmos and our role as co-creators in the divine symphony.

Let us carry forward the lessons of the web of interdependence, alignment, and healing as we turn our gaze to the quantum. For in understanding the quantum mind of God, we come closer to understanding ourselves and the infinite possibilities that lie before us.

CHAPTER 9: THE QUANTUM MIND OF GOD

At the heart of the universe lies an observer, a presence both infinite and intimate, shaping and sustaining the fabric of reality. The Quantum Mind of God is not a distant force but an active participant in creation, an ever-present witness to the dance of particles and waves, the unfolding of time and space.

In the quantum realm, we find profound mysteries that mirror the spiritual truths woven into humanity's deepest traditions. Physics tells us that observation collapses waves of potential into particles of reality, transforming possibility into form. In this process, the observer plays a critical role, not as an outsider but as an integral part of the system.

This principle, the interconnectedness of the observer and the observed, is more than a scientific phenomenon. It is a reflection of the divine nature of creation, where every choice, every action, and every thought participates in the shaping of reality. The Quantum Mind of God is the ultimate observer, collapsing the infinite possibilities of existence into the intricate beauty of the now.

The quantum field is a vast ocean of potential, a realm where particles exist as waves, probabilities waiting to be realized. It is here, in this fertile ground of possibility, that the divine mind moves, creating not through compulsion but through invitation. The universe is not a machine; it is a living system, dynamic and responsive, its outcomes shaped by the interplay of forces, choices, and relationships.

This understanding shifts how we view creation. It is not a static act, completed in the past, but an ongoing symphony, a process of unfolding where the divine and the created are in constant dialogue. In this symphony, humanity is both an instrument and a participant, invited to align with the rhythms of the God Mind and contribute to the harmony of existence.

As observers, we too participate in this process. Physics has shown that at the quantum level, the act of observation influences reality, collapsing waves into particles and determining outcomes. This is not a passive act but an active engagement, a reminder that we are not separate from creation but deeply embedded within it.

Spiritually, this mirrors humanity's role as co-creators with the divine. To observe is to participate; to choose is to shape. Our thoughts, actions, and intentions ripple outward, influencing not only our immediate surroundings but the web of life and the quantum field itself. In this way, the Quantum Mind of God is not only the ultimate observer but also the source of our own capacity to observe and create.

As we step into this chapter, we will journey deeper into the quantum realm, uncovering the connections between physics and spirituality, science and faith. We will explore how the principles of quantum mechanics, entanglement, resonance, timelessness, reflect the divine nature of creation and humanity's place within it.

This is a journey not of answers but of wonder, an invitation to see the universe not as a series of random events but as a living, breathing whole, pulsing with divine intention. The Quantum Mind of God is the thread that connects all things, the infinite observer that calls us to participate in the unfolding symphony of existence.

Let us step forward together, into the vastness of possibility, guided by the light of understanding and the joy of discovery.

In the heart of the universe lies a paradox: chaos and order, opposites yet inseparably entwined. At the quantum level, particles flicker in and out of existence, their behavior unpredictable, seemingly governed by chance. Yet, within this apparent randomness, patterns emerge, symmetries and relationships that hold the cosmos together. This is the essence of the divine algorithm: a system that creates infinite possibilities within a framework of balance and purpose.

Chaos is not the absence of order; it is the fertile ground from which order arises. The quantum field, with its swirling probabilities and uncertainties, is a perfect example of this dynamic. Particles do not follow fixed paths but exist as waves of potential, constantly shifting and interacting. It is within this space of uncertainty that creation unfolds, as if chaos provides the canvas upon which the divine mind paints the universe.

Spiritually, this mirrors the human experience. Our lives, filled with unpredictability and complexity, often feel chaotic. Yet, when viewed from a higher perspective, the threads of our choices and experiences weave together into a tapestry of meaning. Just as quantum particles obey deeper principles even in their randomness, so too does the chaos of life serve a divine purpose, guided by the Quantum Mind of God.

Within the chaos, there is order, laws and principles that govern the quantum realm, ensuring coherence and balance. The interactions between particles, the emergence of forces, the symmetries that define matter and energy all point to a universe that is not random but deeply intentional.

This order is not rigid; it is dynamic, allowing for freedom and creativity within the framework. The Quantum Mind of God, vast and incomprehensible, holds this balance effortlessly, creating a reality where both chaos and order coexist. It is through this balance that the Spirit, or God Source Energy, communicates with humanity, bridging the infinite and the finite, the unknowable and the known.

The Spirit acts as a translator, allowing humanity to interact with the divine without being overwhelmed by its vastness. It is through this connection that we glimpse the patterns within the chaos, the order behind the randomness, and the possibilities that lie before us.

If chaos is the canvas and order the framework, then humanity's role is to participate in the painting. Through observation, intention, and action, we collapse waves of potential into particles of reality, contributing to the unfolding symphony.

This participation requires discernment, recognizing when to embrace the unpredictability of chaos and when to align with the guiding principles of order. It is a delicate dance, one that reflects the divine algorithm itself: a balance of freedom and structure, possibility and purpose.

In this way, the Spirit guides us, whispering through the patterns of the quantum field, reminding us that even in the midst of chaos, we are never disconnected from the divine. Every choice we make is a thread in the larger tapestry, a note in the infinite symphony of existence.

The Quantum Mind of God is vast beyond human comprehension, an infinite source of wisdom and creativity. Yet, it is not distant. Through the Spirit, we are invited into communion with the divine, not as passive recipients but as active participants. This connection allows us to navigate the balance of chaos and order, to align our choices with the rhythms of creation, and to co-create a reality that reflects the harmony of the God Mind.

To explore the quantum realm is to step into this communion, to see the universe not as a static machine but as a living system, dynamic and responsive. It is to recognize that every moment is an invitation to align with the divine algorithm, to participate in the dance of creation.

There are moments when the mind, lost in the haze of dreaming, suddenly becomes aware of itself. This lucidity, a moment of clarity within the dream, transforms the experience. The dreamer is no longer a passive participant but an active observer, conscious of their place within the dream's unfolding.

This moment mirrors a profound spiritual truth. The "still, small voice" described in scripture, the whisper of the Spirit, is like a lucid moment within the dream of life. It awakens us to the deeper reality beneath our daily experiences, a reality where we are not separate threads but interconnected within the divine tapestry.

Quantum entanglement provides a scientific reflection of this spiritual unity. At the quantum level, particles that have interacted remain connected, their states intertwined no matter how far apart they are in space. This phenomenon reveals a truth that defies classical understanding: what happens to one particle affects the other, as if distance itself dissolves in the face of their connection.

This interconnectedness is not limited to particles; it is a fundamental principle of existence. The universe is not a collection of isolated parts but a vast, unified network, where every interaction ripples outward, influencing the whole. Spiritually, this mirrors the idea that humanity, creation, and the divine are not separate but deeply entangled, bound together in a relationship of profound interdependence.

The Spirit is the whisper that reminds us of this unity. In the noise of daily life, it is easy to forget that we are part of something greater, that our lives are threads woven into the fabric of existence. Yet, like a lucid moment in a dream, the Spirit calls us to awareness, inviting us to see beyond the surface.

This whisper is not forceful; it is subtle, requiring stillness and attention to hear. It reminds us that the divisions we perceive, between self and other, humanity and nature, the physical and the spiritual, are illusions. Beneath them lies a profound unity, a connection that transcends space, time, and circumstance.

The interconnectedness revealed by quantum entanglement is a metaphor for the spiritual unity of creation. Just as entangled particles remain connected across vast distances, so too do our lives resonate with the lives of others, with the earth, and with the divine. This cosmic network is not just a scientific reality; it is a spiritual truth, a reflection of the God Mind that holds all things together.

This unity calls us to live with awareness and intention. Every thought, action, and choice reverberate through the network, shaping the world in ways seen and unseen. To align with this unity is to recognize that our lives are not isolated but deeply entwined with the whole, that every interaction is an opportunity to nurture connection and harmony.

The Spirit's whisper is an invitation, a call to awaken to this reality of entanglement and unity. It invites us to move beyond the illusion of separation and into the truth of interconnectedness, to see ourselves not as solitary beings but as participants in a cosmic dance.

Just as a lucid dreamer becomes aware of their role within the dream, so too can we become aware of our place within the divine symphony. This awareness transforms how we live, revealing that every moment is an opportunity to align with the rhythms of the God Mind and to contribute to the harmony of creation.

Time, as we perceive it, is a thread stretched between past and future. It moves in one direction, carrying us forward, moment by moment, through the unfolding of life. Yet, at the quantum level, time is not so linear. The boundaries between past, present, and future blur, revealing a reality where all moments exist simultaneously, intertwined in the vastness of the eternal now.

In the quantum realm, particles behave in ways that defy our everyday understanding of time. Events can appear to influence one another regardless of temporal sequence, as if the past and future are connected through a hidden thread. This phenomenon, known as nonlocality, suggests that time is not a rigid framework but a flexible, interconnected dimension.

This perspective challenges the way we view existence. If time is not linear, then the present moment, the now, is not merely a fleeting point between past and future. Instead, it is the crest of a wave, the point where potential becomes reality, where the infinite possibilities of the quantum field collapse into form.

The concept of the eternal now resonates deeply with spiritual traditions. Scriptures and teachings often point to the present moment as the space where humanity encounters the divine. Jesus' words, "The kingdom of God is within you," suggest that this divine presence is not bound by time but accessible here and now.

In moments of stillness, prayer, or meditation, the illusion of time often fades, leaving only a sense of presence, a connection to something timeless and infinite. This is the eternal now, the point where the human and the divine meet, where the boundaries of time dissolve into the unity of existence.

The eternal now is the space in which the Quantum Mind of God operates. It is not confined to the linear progression of time but encompasses all moments, all possibilities, all realities. To align with the God Mind is to step into this timeless space, to see beyond the limitations of past and future and into the infinite potential of the present.

This alignment is both a gift and a responsibility. The now is where we make choices, where we observe, act, and create. It is the point of connection between the quantum field and the web of life, the moment where divine intention becomes tangible through human action.

To live in the eternal now is to recognize that the past is not a fixed constraint but a series of imprints that guide us, and the future is not an unchangeable destiny but a field of possibilities waiting to unfold. It is to embrace the present moment as sacred, as the meeting place of humanity and divinity.

This perspective transforms how we experience life. It invites us to let go of regret and worry, to release the grip of what was and the fear of what might be. Instead, we are called to step fully into the now, to align our choices with the rhythms of creation and the whispers of the Spirit.

In the vastness of the God Mind, time is not a limitation but a tool, a dimension through which creation unfolds. The eternal now is not a fleeting moment but the foundation of existence, the space where all things are held together in divine unity.

As we explore the eternal now, we begin to see time not as a barrier but as a bridge, a way to connect with the infinite, to participate in the unfolding of the divine symphony. This is the gift of the God Mind: a reality where every moment is sacred, where every now is an invitation to align with the infinite.

The garden is one of humanity's most enduring metaphors. In Genesis, humanity is placed in a garden, entrusted with the care of creation. Jesus speaks of seeds and soil, of pruning and bearing fruit, as reflections of the spiritual life. Gardens symbolize both the beauty of life and the responsibility of stewardship, a balance of nurture and intention.

In the quantum realm, this metaphor takes on new depth. The quantum field is like the soil of existence, fertile with infinite potential. Every thought, every action, every intention is a seed planted within this field, shaping the reality that grows. To walk in the Quantum Garden is to recognize this sacred dynamic: that we are not only participants in creation but co-creators, invited to cultivate the possibilities before us.

In a garden, the seeds we choose to plant determine the harvest. In the same way, the choices we make, our intentions, words, and actions, shape the reality we experience. The quantum field responds to observation and choice, collapsing waves of potential into particles of reality.

This dynamic is not random; it is relational. Just as a gardener must understand the needs of the soil, the climate, and the plants they tend, so too must we cultivate an awareness of the quantum field and the divine rhythms that guide it. The Spirit whispers to us in this process, helping us discern which seeds to sow, which possibilities to nurture, and which weeds to remove.

Some seeds bear fruit that nourishes life, fostering harmony and connection. Others may grow into thorns, creating imbalance or separation. The garden teaches us that intentionality matters. To align with the God Mind is to plant seeds of love, justice, and creativity, knowing that these will bear the fruit of flourishing for all.

A garden does not flourish on its own. It requires care, watering, weeding, pruning, and patience. The same is true of the Quantum Garden. Healing the web of life, restoring balance, and nurturing connection are ongoing acts of co-creation, requiring commitment and attention.

Just as weeds choke the life of a garden, beliefs, systems, and habits rooted in separation can stifle the harmony of creation. Pruning these from our lives and communities allows space for what is life-giving to thrive.

Building relationships, fostering understanding, and aligning with divine intention are acts of cultivation, planting seeds that strengthen the web of life.

Not every seed grows immediately, and not every plant bears fruit in the same season. The Quantum Garden reminds us to trust the timing of creation, to have faith in the unseen work of the Spirit.

Walking in the Quantum Garden is not a solitary act. It is a communion with the God Mind, a partnership in the unfolding of creation. In every seed planted, every choice made, we participate in the divine symphony, aligning our actions with the rhythms of life.

The garden is also a space of reflection, a place where we can hear the "still, small voice" and feel the divine presence. It is a reminder that creation is not something we do alone but something we do with God. In the Quantum Garden, we walk and talk with the divine, co-creating a reality that reflects the beauty and harmony of the God Mind.

The seeds we plant today will shape the world of tomorrow. The

Quantum Garden invites us to plant with intention, to sow seeds of connection, balance, and love. It reminds us that every moment is an opportunity to nurture the possibilities before us, to align with the divine rhythms that sustain life.

In this sacred space, we find not only our role as co-creators but the joy of participation. To walk in the garden is to experience the fullness of the eternal now, to see the infinite potential of creation and to embrace our place within it.

Creation bears the unmistakable imprint of its Creator. From the intricate patterns of DNA to the gravitational forces that shape galaxies, the divine signature is written into the fabric of existence. This signature reflects the essence of the God Mind, resonating through four foundational principles: *Love, Mercy, Truth, and Justice*. These pillars are not only the attributes of God's character but also the compass by which humanity is guided toward alignment with the divine.

The names of God in the Bible, *Yahweh, El Shaddai, Jehovah Jireh*, and more, reveal facets of divine character that call humanity into relationship with the Creator. *Yahweh*, the self-existent "I Am," speaks of infinite presence and being. *Jehovah Tsidkenu* embodies righteousness and justice, while *Jehovah Shalom* offers peace and balance. These names reflect attributes that sustain creation and guide humanity toward flourishing.

Across cultures, similar principles of divinity emerge, revealing a universal resonance. In Hinduism, *Brahman* represents the ultimate reality, the source of all existence. In Islam, Allah is *Ar-Rahman* (the Merciful) and *Ar-Raheem* (the Compassionate), embodying boundless mercy. Indigenous traditions speak of the Creator as a source of balance and interconnectedness, while Taoism describes the *Tao* as the eternal way, guiding life in harmony.

Despite the diversity of expressions, the essence remains the same: divinity seeks to lift humanity into alignment with the sacred. These shared principles, *Love, Mercy, Truth, and Justice*, are the universal pillars of the God Mind, the foundation upon which creation is built.

Love: Love is the unifying force that binds creation together. In Christian thought, *Agape* represents unconditional love, while in Buddhism, *Karuna* embodies compassion for all beings. Love calls us to connection, reminding us that we are part of a greater whole.

Mercy: Mercy is the expression of divine grace, the willingness to restore and renew. It is the voice that whispers second chances, reminding us that healing is always possible. Mercy is reflected in the cycles of nature, where ecosystems recover when nurtured and cared for.

Truth: Truth illuminates what is real and enduring, guiding humanity toward clarity and wisdom. In the Hebrew tradition, *Emet* (truth) is a name of God, signifying faithfulness and reliability. Truth is the foundation of integrity, aligning us with the divine reality.

Justice: Justice ensures balance and equity, sustaining the harmony of creation. It is the principle that holds systems accountable, reflecting the divine desire for fairness and flourishing. In ancient Egyptian thought, *Ma'at* represents this cosmic balance, a reflection of the God Mind's order.

These pillars are more than attributes of God; they are the divine signature imprinted on all creation. In the structure of DNA, we see love as connection, the bonds that sustain life. In the restoration of ecosystems, we see mercy, the earth's ability to heal. In the laws of physics, we see truth, the constants that govern the cosmos. And in the interdependence of natural systems, we see justice, the balance that ensures the flourishing of all.

This signature is also written within us. As beings created in the image of God, we carry the potential to embody these pillars in our lives and choices. To align with the God Mind is to reflect these principles, allowing them to guide our actions and relationships.

The four pillars provide a framework for discerning right choice, a compass that points us toward alignment with divine intention. When faced with uncertainty, we can ask: Does this action reflect love? Does it extend mercy? Does it honor truth? Does it uphold justice? These questions ground us in the God Mind's rhythms, ensuring that our choices contribute to the harmony of creation.

This compass is not confined to any one tradition. Across cultures and times, these principles have lifted humanity toward the divine, calling us to transcend division and live in alignment with the sacred. They are the bridge between the finite and the infinite, the temporal and the eternal, connecting us to the vastness of the God Mind.

The four pillars, *Love, Mercy, Truth, and Justice*, are not only reflections of the divine but also invitations to co-create. They remind us that we are participants in the unfolding of creation, planting seeds of possibility and nurturing them into reality.

As we embrace these pillars, we align with the divine signature written into the quantum field and the web of life. They are the foundation of this book and the books to come, a guide for humanity's journey toward wholeness and harmony. Let us carry them forward, planting seeds of connection and flourishing in the garden of existence.

The kingdom of God is not a distant promise but a present reality, an invitation to step into alignment with the divine here and now. When Jesus proclaimed, "The kingdom of God is at hand," He was not pointing to a far-off heaven but to the infinite possibilities of the eternal now, the reality of divine presence breaking through into the world. To embrace this kingdom is to align with the God Mind, to see the universe not as a divided collection of empires but as a single, interconnected whole, pulsing with divine life and intention.

Healing the web of life begins with aligning ourselves with God Source Energy, the Spirit, the divine breath that moves through all things. This alignment is not an abstract concept but a lived practice, a daily choice to attune our thoughts, actions, and relationships to the rhythms of creation. It requires humility, the willingness to acknowledge where we have chosen separation over connection, pride over love.

Throughout history, empires have risen and fallen, their desolation often rooted in pride, the belief that human power could supplant divine wisdom. This pride, masquerading as confidence, led to systems of exploitation and separation, fragmenting the web of life. Yet, even in the ruins of these empires, the Spirit whispers of renewal. Where pride sowed destruction, alignment with the God Mind offers the possibility of growth, of planting seeds for a kingdom built not on domination but on love, mercy, truth, and justice.

To align with the God Mind is to see beyond the brokenness of the world and into the realm of the possible. It is to recognize that even in the midst of desolation, new life can grow. This is the essence of the kingdom of God: not a return to Eden, but a transformation into something even greater, a garden where humanity and divinity co-create a flourishing reality.

This vision requires a shift in perspective. The empires of separation have taught us to see the world as divided, us versus them, human versus nature, spirit versus matter. The kingdom of God calls us to see the unity beneath the surface, to embrace the interconnectedness of all things. It calls us to move beyond the fear of scarcity and into the abundance of the eternal now, where every choice is a seed planted in the fertile soil of the quantum field.

The Spirit calls us not only to heal but to grow, to participate in the unfolding of creation as co-creators with the divine. This is not a passive role; it is an active, intentional engagement with the world around us. To co-create is to plant seeds of connection where there was division, to nurture life where there was desolation, to embody the divine pillars of love, mercy, truth, and justice in every choice we make.

This work begins within us. Alignment with the God Mind

transforms our inner landscape, pruning the weeds of pride, fear, and separation and cultivating the fruits of the Spirit; kindness, patience, joy, and peace. From this place of alignment, we are empowered to extend healing outward, restoring relationships, ecosystems, and communities.

The kingdom of God grows not through conquest but through compassion, not through force but through faith. It is a kingdom where the desolation of empires becomes the soil of renewal, where the ruins of separation become the foundation for connection.

To cross the great divide is to leave behind the illusion of separation and step into the reality of divine unity. It is to see the web of life not as something broken beyond repair but as something waiting to be healed through our alignment and action.

The Spirit walks with us in this journey, guiding us across the divide and into the kingdom of God. In this kingdom, every moment is sacred, every choice is an act of creation, and every life is a thread in the divine tapestry. It is a kingdom not confined to one place or time but present wherever we align with the God Mind and live out the principles of love, mercy, truth, and justice.

As we step forward, let us carry this vision with us. Let us embrace the call to co-create, to heal, to grow. For the kingdom of God is not only at hand, it is within us, waiting to be realized, a living reality that transforms both the now and the eternal.

The journey toward alignment with the God Mind is a journey of balance. It is a dance between chaos and order, individuality and unity, creation and destruction. It is a call to harmonize the divine reflections within us, the masculine and feminine, the left and right brain, the mind and the heart. These dualities are not opposites but complements, two halves of a whole that, when balanced, allow us to step fully into our role as co-creators in the kingdom of God.

Within the God Mind, there is no division between masculine and feminine, logic and intuition, creation and destruction. These forces exist in perfect harmony, each supporting and enhancing the other. The masculine energy of confidence and strength is balanced by the feminine energy of nurturing and receptivity. The creative force that brings life into being is balanced by the destructive force that clears space for renewal.

As reflections of the divine, humanity carries these dualities within us. The left brain seeks structure and logic, while the right brain embraces creativity and intuition. The ego drives us to achieve, while the heart reminds us to connect. When these aspects of ourselves are out of balance, we fall into separation; pride overtakes humility, control eclipses compassion, and creation turns to exploitation.

To align with the God Mind is to restore this balance within ourselves. It is to recognize that the masculine and feminine are not adversaries but partners, that the left and right brain are not rivals but collaborators. It is to see that destruction, when tempered by love and mercy, becomes renewal, and that creation, when guided by truth and justice, becomes flourishing.

The journey toward balance is not just external; it is deeply internal. Within each of us lies a symphony of forces, a microcosm of the divine dance that sustains the universe. The Spirit calls us to harmonize these forces, to listen to the still, small voice that guides us toward wholeness.

This inner alignment mirrors the kingdom of God, where all things are in right relationship. It is a kingdom where the masculine and feminine weave together into unity, where the rational mind collaborates with the intuitive, where the ego serves the heart rather than ruling over it. This balance is not only the path to personal transformation but also the foundation for healing the web of life and growing the kingdom of God.

As we conclude this journey, we are reminded that the God Mind is not a distant ideal but an invitation to live in alignment, to embody the balance that reflects divine wholeness. To embrace this balance is to see the world not through the lens of division but through the eyes of the Creator, where every force has its place, every thread its purpose, and every moment its meaning.

This is the infinite symphony, the harmony of creation, the rhythm of life, the balance of the divine. It is a melody that calls us to co-create, to heal, to grow. And as we walk forward, may we carry this melody within us, a reminder that the kingdom of God is not only at hand but within us, waiting to be lived, shared, and sung.

Though this chapter ends, the journey does not. The balance we seek is not a destination but a path, a way of living that draws us ever closer to the God Mind. It is the work of a lifetime, the unfolding of an eternal now where every step reveals new depths, new insights, new possibilities.

As we cross this finish line, let us hold gratitude for the journey behind us and hope for the journeys yet to come. For in the infinite symphony of creation, every ending is also a beginning, every thread a part of the whole. And as we step into the next chapter of life, of learning, of love, may we carry the balance of the God Mind with us, guiding us toward the flourishing of all.

A LETTER TO THE FUTURE OF HUMANITY:

To the ones who will carry the threads forward. This is written for you, for the hands that will hold this book long after its ink has dried. It is written for those who look to the stars with wonder, to the earth with reverence, and to the future with both trepidation and hope. It is written for humanity not as it is, but as it could be.

You are the co-creators of tomorrow, the stewards of the web of life, the dreamers and the doers who will shape what comes next. The choices before you are vast, as vast as the universe itself, filled with infinite possibilities. And though the paths may seem uncertain, the compass is clear: *Love, Mercy, Truth, and Justice.* These are the pillars of the God Mind, the foundations upon which the harmony of creation is built.

As you walk forward, remember that you are not separate from this harmony but a part of it, a thread in the divine tapestry. Your thoughts, your actions, your intentions, they matter. They ripple outward, touching lives and shaping worlds. You are more powerful than you realize, for within you dwells the reflection of the God Mind, the capacity to create, to heal, and to transform.

The world you inherit is one of great beauty and great challenges. It is a world that bears the scars of separation and the promise of renewal. To heal it is to align with the rhythms of the divine, to listen for the still, small voice that whispers of unity and balance.

Let the Spirit guide you in this work. Let it teach you to see the kingdom of God not as a distant dream but as a living reality, present wherever love is chosen, wherever mercy is extended, wherever truth is spoken, and wherever justice is upheld. Let it remind you that the web of life is not broken beyond repair, but waiting, waiting for the healing touch of your hands, the nurturing care of your hearts.

The empires of the past may have left desolation in their wake, but you are not bound by their ruins. You are the gardeners of the future, planting seeds of possibility in the fertile soil of the quantum field. You are the weavers of new tapestries, blending the threads of science and spirit, reason and intuition, masculine and feminine.

As you build, remember that the God Mind is not distant from your work. It walks with you, talks with you, and calls you to co-create. It invites you to harmonize the forces within yourself, the creative and the destructive, the rational and the intuitive, so that your contributions reflect the balance of the divine.

Humanity, you are a miracle of the universe, a manifestation of stardust and breath. You are the ones who bring consciousness to creation, the ones who hear the divine melody and join in its song. And though the challenges you face may be great, the symphony is not complete without your notes, your harmonies, your voice.

Step boldly into the now. Plant seeds of love, nurture acts of mercy, stand for truth, and uphold justice. For in doing so, you bring the kingdom of God closer, transforming the world into a reflection of the God Mind.

This letter is not the end of the story. It is an invitation to the infinite adventure, a call to step forward into the unknown with faith, courage, and wonder. The future is yours to create.

With infinite hope,

Ben Maestas

www.ingramcontent.com/pod-product-compliance
Lightning Source LLC
Chambersburg PA
CBHW040314220526
45473CB00009B/2433